Undergraduate Lecture Notes in Physics

Undergraduate Lecture Notes in Physics (ULNP) publishes authoritative texts covering topics throughout pure and applied physics. Each title in the series is suitable as a basis for undergraduate instruction, typically containing practice problems, worked examples, chapter summaries, and suggestions for further reading.

ULNP titles must provide at least one of the following:

- An exceptionally clear and concise treatment of a standard undergraduate subject.
- A solid undergraduate-level introduction to a graduate, advanced, or non-standard subject.
- A novel perspective or an unusual approach to teaching a subject.

ULNP especially encourages new, original, and idiosyncratic approaches to physics teaching at the undergraduate level.

The purpose of ULNP is to provide intriguing, absorbing books that will continue to be the reader's preferred reference throughout their academic career.

Series editors

Neil Ashby
Professor Emeritus, University of Colorado, Boulder, CO, USA

William Brantley
Professor, Furman University, Greenville, SC, USA

Matthew Deady
Professor, Bard College Physics Program, Annandale-on-Hudson, NY, USA

Michael Fowler
Professor, University of Virginia, Charlottesville, VA, USA

Morten Hjorth-Jensen
Professor, University of Oslo, Oslo, Norway

Michael Inglis
Professor, SUNY Suffolk County Community College, Long Island, NY, USA

Heinz Klose
Professor Emeritus, Humboldt University Berlin, Berlin, Germany

Helmy Sherif
Professor, University of Alberta, Edmonton, AB, Canada

More information about this series at http://www.springer.com/series/8917

Enzo De Sanctis · Stefano Monti
Marco Ripani

Energy from Nuclear Fission

An Introduction

 Springer

Enzo De Sanctis
Laboratori Nazionali di Frascati
Istituto Nazionale di Fisica Nucleare (INFN)
Frascati (Rome)
Italy

Marco Ripani
Istituto Nazionale di Fisica
  Nucleare (INFN) - Genova
Genova
Italy

Stefano Monti
International Atomic Energy Agency
Vienna International Centre
Vienna
Austria

ISSN 2192-4791   ISSN 2192-4805   (electronic)
Undergraduate Lecture Notes in Physics
ISBN 978-3-319-30649-0   ISBN 978-3-319-30651-3   (eBook)
DOI 10.1007/978-3-319-30651-3

Library of Congress Control Number: 2016937961

Printed on acid-free paper

This Springer imprint is published by Springer Nature
The registered company is Springer International Publishing AG Switzerland

The original version of the chapters was revised: The figures and inline equation have been replaced in Chapters 2 and 3.
The erratum to the chapters is available at: 10.1007/978-3-319-30651-3_7

Preface

The main goal of this book is to provide an accessible and concise introduction to nuclear physics and energy production from nuclear fission. The book conveys quantitative information on the fundamental physical mechanisms at play in the nuclear fission process, as well as on current exploitation of fission in power plants, including technical aspects and trends. It provides a basis for a core undergraduate course in this area.

The text is divided into two parts; the first part encompasses the basics of nuclear forces and properties of nuclei, nuclear collisions, nuclear stability, radioactivity, as well as a detailed discussion of the fission process and relevant topics in its application to energy production. The second part covers the basic aspects of nuclear reactor technologies, nuclear fuel cycle and resources, reactor safety and regulation, security and safeguards, and spent fuel and radioactive waste management. The book also contains some sections devoted to a qualitative description of the phenomena associated with the passage of charged particles and radiation through matter, a discussion of the biological effects of nuclear radiation and of radiation protection, and a summary of the ten most relevant accidents occurred to nuclear installations, some of which have had a significant impact on the development and deployment of nuclear power. A glossary at the end of the book provides a handy reference to the terminology used in nuclear physics and nuclear energy.

The subject matter is broad and somewhat heterogeneous and obviously does not allow going deep into the subtleties of each single topic. However, the aim is to provide an as complete as possible overview of the many aspects and issues involved in the deployment of nuclear power. In topics ranging from the fundamental physical principles to the much-debated challenges of safety and closure of the nuclear cycle, whenever possible, authoritative sources of information are used (typically international agencies and institutions), thereby stimulating the reader to expand his/her knowledge on each topic by looking at the suggested references, or by searching further technical literature on the Web.

The book is suitable for undergraduates in physics, nuclear engineering, and other science subjects. However, mathematics is kept at a level that can be easily followed by a wide audience. The addition of solved problems, strategically placed throughout the text, and the collections of problems at the end of the chapters help to better understand the scientific and technical topics presented in the text, and allow appreciating the quantitative aspects of various phenomena and processes. Many illustrations and graphs effectively supplement the text and help visualising specific issues.

Contents

Part II   Energy from Nuclear Fission

Abbreviations

ABWR	Advanced Boiling Water Reactor
ACR	Advanced CANDU Reactor
ADS	Accelerator-Driven System
AGR	Advanced Gas-cooled Reactor
ALARA	As Low As Reasonably Achievable
ALARP	As Low As Reasonably Practicable
APWR	Advanced Pressurised Water Reactor
BDBA	Beyond Design-Basis Accidents
BSS	Basic Safety Standards
BWR	Boiling Water Reactor
CANDU	CANadian Deuterium Uranium reactor (PHWR type)
CCS	Carbon Capture and Storage
CDF	Core Damage Frequency
CT	Computed Tomography
CTBT	Comprehensive Nuclear Test Ban Treaty
DBA	Design Basis Accident
DNA	Deoxyribonucleic Acid
DSA	Deterministic Safety Approach
DT	Doubling Time
ECCS	Emergency Core Cooling System
EDG	Emergency Diesel Generator
EPR	Evolutionary Pressurised Reactor
ESBWR	Economic Simplified Boiling Water Reactor
FBR	Fast Breeder Reactor
GCR	Gas-Cooled Reactor
GFR	Gas-cooled Fast Reactor
GIF	Generation IV International Forum
HEU	Highly Enriched Uranium
HLW	High-Level Waste
HTGR	High-Temperature Gas-cooled Reactor
HTR	High-Temperature Reactor

HWR	Heavy-Water Reactor
IAEA	International Atomic Energy Agency
ICRP	International Commission on Radiological Protection
IEA	International Energy Agency
ILW	Intermediate-Level Waste
INES	International Nuclear Event Scale
ITER	International Thermonuclear Experimental Reactor
JET	Joint European Torus
LEU	Low-Enriched Uranium
LFR	Lead-cooled Fast Reactor
LLW	Low-Level Waste
LMFBR	Liquid Metal Fast Breeder Reactor
LNT	Linear, No-Threshold
LOCA	Loss-Of-Coolant Accident
LTO	Long-Term Operation
LWGR	Light Water Graphite Reactor
LWR	Light Water Reactor
MOX	Mixed-Oxide Fuel
MSBR	Molten Salt Breeder Reactor
MSR	Molten Salt Reactor
NEA	Nuclear Energy Agency (OECD)
NORM	Naturally Occurring Radioactive Materials
NPT	Treaty on the Non-Proliferation of Nuclear Weapons
O&M	Operation and Maintenance
OECD	Organisation for Economic Co-operation and Development
P&T	Partitioning and Transmutation
PBMR	Pebble Bed Modular Reactor
PET	Positron Emission Tomography
PGAA	Prompt Gamma Activation Analysis
PHWR	Pressurised Heavy Water Reactor
PIXE	Proton Induced X-ray Emission
PSA	Probabilistic Safety Assessment
PUREX	Plutonium Uranium Reduction Extraction
PWR	Pressurised Water Reactor
R&D	Research and Development
RAR	Reasonably Assured Resource
RBMK	Russian Abbreviation for Graphite-moderated Light Water-cooled Reactors
RBS	Rutherford Back Scattering
RDT	Reactor Doubling Time
RNA	Ribonucleic Acid
SCRAM	Safety Control Rod Axe Man
SCWR	Supercritical-Water-cooled Reactor
SFR	Sodium-cooled Fast Reactor
SMR	Small Modular Reactor

SNF	Spent Nuclear Fuel
SPECT	Single-Photon Emission Computed Tomography
TMI	Three Mile Island
UNSCEAR	United Nations Scientific Committee on the Effects of Atomic Radiation
VHTR	Very-High-Temperature Reactor
VVER	Russian Design of Pressurised Water Reactor
WANO	World Association of Nuclear Operators
WENRA	Western European Nuclear Regulators' Association
WIPP	Waste Isolation Pilot Plant (United States)
WNA	World Nuclear Association

Part I
Nuclear Physics and Radioactivity

Chapter 1
The Building Blocks of Matter

After a short excursus on the atom, the basic building block of matter, the Chapter deals with the most basic properties of nuclei, their composition, size, mass and density. The nucleus is a tiny bundle of concentrated mass and energy at the centre of the atom. It is about one hundred thousand times smaller than the atom itself and is, in turn, an aggregate of other particles: protons and neutrons.

The Chapter then describes the nuclear force responsible for binding protons and neutrons, the nuclear binding energy, the diagram of known nuclei and the valley of stability, to conclude with a brief discussion on nuclear reactions and the abundances of elements in the Universe.

1.1   The Atom and Its Constituents

All matter around us, both living and inanimate, is made of atoms. They are the basic building blocks of all structures and organisms in the universe. The stars, the Sun, the planets, the stones, the trees, and the air we breathe, the animals and the human beings are all made of different combinations of atoms.

Atoms are very small compared to common objects; their size is of the order of 10^{-10} m (a ten-millionth of a millimetre). A hundred million atoms, side by side, are required to form a line one centimetre long (see Fig. 1.1a). However, they are very numerous: in the head of a pin, which has a mass of only about 8 mg, there are around 10^{20} (one hundred billion of billions) iron atoms. Lining up, side by side, the atoms contained in one mole[1] of any substance, one could cover a distance 400 times

---

[1]The mole of an element is the quantity of the element that contains as many atoms as there are in 12 g of the isotope carbon-12 (^{12}C); thus, by definition, one mole of pure ^{12}C has a mass of exactly 12 g. This number, which is called *Avogadro number* N_A, has the value $6.02214179 \times 10^{23}$ atoms of the element. The mole is one of the base units of the International System of Units; it has the unit symbol mol.

© Springer International Publishing Switzerland 2016
E. De Sanctis et al., *Energy from Nuclear Fission*, Undergraduate Lecture
Notes in Physics, DOI 10.1007/978-3-319-30651-3_1

(a)

(b)

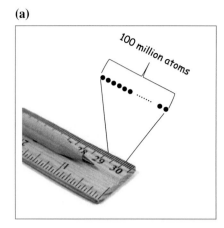

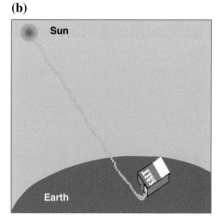

Fig. 1.1 **a** Atoms have a very small size with diameter of a few tenth of nanometre. In order to form a segment 1 cm length, one should put in row one hundred million of them. **b** Atoms are very numerous; lining up side by side the atoms forming a grain of salt, one could cover a distance much greater than that between the Earth and the Sun, which is about 150 million kilometres

greater than that between the Earth and the Sun (about 1.5×10^{11} m, or 150 million kilometres) (see Fig. 1.1b).

The two Greek philosophers Democritus and Leucippus, in the fifth century BC, were the first to introduce the idea that everything in the Universe is composed of indivisible myriads of tiny indivisible particles, which they called *atoms*, which in ancient Greek meant: "it cannot be split into smaller pieces". Contrary to their thought, the atom is actually a complex system that can be broken into pieces.

In fact, atoms are formed by a central core, the nucleus, with positive electric charge, surrounded by distributions of negative electric charges, the electrons. The number of electrons of an atom is called *the atomic number* of the atom and is always denoted by Z.

The nucleus contains over 99.95 % of the mass of the atom, but occupies only a very tiny fraction of its volume, about 10^{-15} (one millionth of a billionth). The electrons contain the remaining mass and move in the much larger atomic volume. In the lightest atom, hydrogen formed by 1 proton and 1 electron, the nucleus has a mass about 1836 times larger than that of the electron, precisely 1.67262×10^{-27} kg against 9.10938×10^{-31} kg. In the heaviest natural atom, uranium formed by 92 protons, 146 neutrons and 92 electrons, the nucleus has a mass about 4750 times larger than that of the 92 electrons.

Atoms are mostly empty space, as nuclei have sizes of the order of 10^{-15} m, while the electronic distributions are relatively very large (size 10^{-10} m). Keeping these proportions, if the nucleus was the size of an ant in the middle of a football stadium, the distributions of electrons would extend as far as the stands. This is pictorially shown in Fig. 1.2, where electrons are shown on defined orbits for illustrative purpose only. Instead, they move around the nucleus along all imaginable paths with different probabilities (see the Insight: Viewing the Atoms).

Fig. 1.2 Atoms are mostly empty space. If the nucleus were the size of an ant in the middle of a football stadium, the distributions of electrons would extend as far as the stands. Here electrons are shown on defined orbits for illustrative purpose only [1]

The nucleus and the electrons in the atom are bound together because, carrying electric charges of opposite sign, they attract each other. However, the total electric charge of the atom is zero, as the total negative charge of the electrons is equal to the positive charge of the nucleus, and charges of opposite sign cancel each other.

An atom can easily loose one or more electrons or, in contrast, can gain one or more electrons: in which case, it is said that the atom is positively ionised, one or more times, or negatively ionised, one or more times, respectively.

Atoms interact with other atoms of the same or different kind, forming larger structures called molecules. The chemical properties of the atoms, that is the degree of their affinity[2] for other atoms, are determined by the number of electrons.

Some chemical transformations occur spontaneously. Two different substances that come close to each other may combine giving origin to a new substance. In these cases, energy is released, usually in the form of thermal energy. In other cases, however, for the transformation to occur, it is necessary to supply energy, for example by heating the container in which the reactants are placed.

The chemical reactions involve the outer electrons of the atoms and electrical forces, and have no chance to modify the structure of nuclei, because the energies involved are vastly lower than those needed to induce nuclear modifications.

1.1.1 Insight: Viewing the Atoms

Atoms and sub-atomic particles behave in peculiar ways that cannot be described by the classical laws of physics. Instead, they obey the laws of quantum mechanics, which have a radically different behaviour from those of the world we see with our eyes.

---

[2]Chemical affinity is the electronic property by which dissimilar chemical species are capable of forming chemical compounds. Chemical affinity can also refer to the tendency of an atom or compound to combine by chemical reaction with atoms or compounds of unlike composition.

(a)   **(b)**

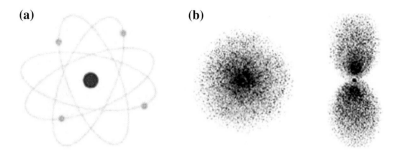

Fig. 1.3 a Outdated schematic representation of an atom with the nucleus at the centre and electrons orbiting around it. Electrons instead can be thought of as buzzing around the nucleus like a swarm of bees. **b** Probability clouds for the two lowest-energy states of the electron in a hydrogen atom: the *large dot* at the centre of the two distributions, spherical on the *left*, and with two vertical lobes, on the *right*, is the nucleus. The greater density of the *small dots* means a higher probability of finding the electron at that particular location

According to these laws, we cannot say exactly where an electron is in an atom; we can only say what the probability is to find it in a given position. Therefore, the view of the atomic structure such as that of a miniaturized macroscopic system is not supported by the modern knowledge of quantum mechanics. For instance, the Bohr-Sommerfeld model,[3] where the atom is depicted as a miniature solar system with the nucleus at the centre and the electrons revolving around it in precisely defined orbits, leads to a wrong view of the situation and has now only historical interest. Nevertheless pictures of the atoms inspired by this old-fashioned model, like the one shown in Fig. 1.3a, are still found today in textbooks. Electrons instead can be thought of as buzzing around the nucleus like a swarm of bees.

Hence, at the atomic scale, one cannot speak of electron orbits. The electrons, instead, can be found anywhere around the nucleus, even very close to and very far from it. However, the probabilities of finding the electron at different distances from the nucleus are different. We must imagine a cloud of electrons (even if the electron is only one as in the case of hydrogen) around the nucleus, which is denser where the electron is most likely to be found, and thinner in regions where the electron is less likely to be found.

In fact, each atomic species has a discrete set of possible configurations, called states, each with a defined energy and each with a certain shape of the cloud of electrons. Quantum mechanics allows for the exact calculation of these energy and probability density values.

---

[3]Niels Henrik David Bohr (Copenhagen, Denmark, 1885—Carlsberg, Denmark, 1962) was a Danish physicist who made foundational contributions to understanding atomic structure and quantum theory, for which he received the Nobel Prize in Physics in 1922. Arnold Johannes Wilhelm Sommerfeld, (Königsberg, Germany, 1868—Munich, Germany, 1951) was a German theoretical physicist who pioneered developments in atomic and quantum physics.

Consider, by analogy, a guitar string. Particularly important are its vibrations at defined frequencies, the fundamental frequency and its harmonics. It is a discrete sequence of vibrational modes with frequencies increasing in harmonic succession (proportionally to the integer numbers). Each vibration mode has a definite shape, each with a growing number of nodes and antinodes. Another analogy can be found in the vibration modes at defined frequency of two-dimensional objects, such as the membrane of a drum, or of three-dimensional objects, such as a bell. Each of these modes can be identified by an integer number—the order of the harmonic—in the case of the string, and by two or three integer numbers if the dimensions are two or three.

The situation is quite similar for atoms, where we can think of the energy of the atomic state as the analogue of the vibration frequency of the string and we can think of the cloud of probability as the analogue of the distribution of the vibrational energy along the string.

As an example, Fig. 1.3b shows the probability clouds for the two lowest-energy states of the electron in a hydrogen atom. The large dot at the centre of the two distributions, spherical on the left, and with two vertical lobes on the right, is the nucleus. The greater density of the small dots means a higher probability of finding the electron at that particular location. The clouds are denser where the electron is most likely to be found. They also indicate the size of the atom in different states; in quantum mechanics, objects do not have a precise and well-defined size, just as a cloud does not have a distinct boundary.

1.2   The Nucleus

The nucleus is the massive centre of the atom in which almost all of the atomic mass is concentrated. It is about one hundred thousand times smaller than the atom itself and is, in turn, an aggregate of other particles: protons and neutrons. In the literature, these particles are denoted by the letters p and n, respectively. Neutrons and protons, in turn, are not elementary particles. They are actually made up of even smaller particles called *quarks*, which are held together by other particles called *gluons*.

Protons and neutrons have similar masses ($M_p = 1.67262 \times 10^{-27}$ kg and $M_n = 1.67493 \times 10^{-27}$ kg, respectively), with the mass of the neutron being greater than that of the proton only by less than 0.14 %. Protons have a positive electric charge, $+e$, of the same value, but opposite in sign, as the negative charge of electrons ($e = 1.6021762 \times 10^{-19}$ C). Neutrons have no electric charge. The total charge of the nucleus is then $+Ze$.

Apart from the electric charge, protons and neutrons have almost the same properties and therefore, very often, the term *nucleon* is used to indicate both.

Each atom contains an equal number of protons and electrons and is identified by two numbers: the atomic number Z, equal to the number of protons (and electrons) it contains, and the mass number A, equal to the total number of nucleons (protons plus neutrons) in its nucleus. The number N of neutrons is also used. The three numbers are connected by the relation:

$$A = Z + N. \tag{1.1}$$

Table 1.1 shows the values of the electric charge and the mass of the electron and the nucleons. The masses are given in three units: kilogram (kg), atomic mass unit (u), and energy unit (eV/c^2); (the latter two units will be defined in Sects. 1.6 and 1.7, respectively). This because the familiar International System (SI) units based on the metre, kilogram, and second are not the most convenient ones when dealing with nuclear dimensions of the order of 10^{-15} m and nuclear masses of the order of 10^{-27} kg. Throughout this book we will generally use the energy unit since energy is a more general concept than mass and is hence more practical in calculations involving nuclear reactions.

It is interesting to note that the proton and the neutron are much heavier than the electron, about 1836 and 1839 times, respectively. Keeping these proportions, if an electron weighed the same as a 1-cent euro coin (2.30 g) or a dime (2.268 g), a proton would weigh about the same as four litres of milk (around 4.2 kg).

All nuclei with the same values of Z and A (and therefore also of N) constitute a particular nuclear species or *nuclide*. A particular nuclide is denoted by adding to the chemical symbol of the corresponding element, X, the mass number, as superscript index, and the atomic number, as subscript: $^{A}X_{Z}$. Thus, $^{15}N_{7}$ denotes the nitrogen nucleus that contains seven protons and eight neutrons, for a total of 15 nucleons.

Since an atom is electrically neutral, the number of protons in a nucleus determines that of electrons in the atom, and then the chemical properties of the atom itself. Therefore, it is redundant to specify both the symbol of the element and its atomic number. In fact, the atomic number Z identifies the chemical element: carbon, oxygen, gold, and so on. For example, if one speaks of the nitrogen nucleus, we know automatically that $Z = 7$. For this reason, for example, one may avoid to write the subscript 7 to $^{15}N_{7}$ and simply write ^{15}N, which one can read "nitrogen fifteen".

Removing or adding one proton in a nucleus transforms an element into another. If one adds or removes one neutron, the number of peripheral electrons remains the same and therefore the chemical properties of the atom are not changed. This means that such an atom should remain in the same place of the Periodic Table of the Elements (see next Section).

Table 1.1 Electric charge and mass of atomic particles [2]

Particle	Symbol	Electric charge	Mass		
		[C]	[kg]	[u]	[MeV/c^2]
Electron	e	$-1.60217662 \times 10^{-19}$	9.10938×10^{-31}	5.485799×10^{-4}	0.511099
Proton	p	$+1.60217662 \times 10^{-19}$	1.67262×10^{-27}	1.007276	938.272
Neutron	n	0	1.67493×10^{-27}	1.008665	939.565

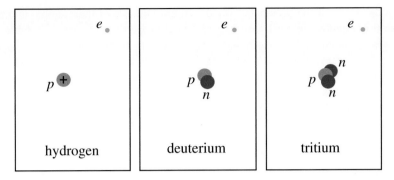

Fig. 1.4 Hydrogen isotopes. The nuclei of the three isotopes with $Z = 1$ have special names: proton, deuteron and triton. The corresponding atoms are named hydrogen, deuterium and tritium

Nuclides with the same number of protons and a different number of neutrons are called *isotopes*. Nuclides with the same number of neutrons and a different number of protons are called *isotones*. Nuclides with the same mass number A, such as ^{3}He and ^{3}H (the former is a lighter helium isotope with two protons and only one neutron, the latter is a hydrogen isotope with one proton and two neutrons, called *tritium*) are called *isobars*. Because of the similarity of the nuclear interactions of protons and neutrons, different isobars have similar nuclear properties.

Nuclide isotopes form atoms with identical chemical properties, but different nuclear properties. For example, we know 6 carbon isotopes, having mass number between 11 and 16; and 14 uranium isotopes, having mass numbers between 227 and 240.

The isotopes of an element are not all present with the same abundance on Earth. For example, in addition to hydrogen, ^{1}H, with a single proton, there are two other isotopes, deuterium (^{2}H) and tritium (^{3}H), whose nuclei contain one and two additional neutrons, respectively (see Fig. 1.4). The nuclei of these three isotopes have special names: proton, deuteron (d) and triton (t). Deuterium, which is stable, is present in nature in the ratio of 1 to 6,420 with respect to hydrogen. This means that in any compound containing a hydrogen atom in its molecule, about every six thousand four hundred molecules there is a deuteron nucleus instead of a proton.

Uranium, which has atomic number $Z = 92$, is found in nature as a mixture of the three isotopes ^{238}U (99.275 %), ^{235}U (0.720 %) and ^{234}U (0.005 %), where numbers in parentheses show the relative percentages.

1.3   The Periodic Table of the Elements

There are 98 naturally occurring kinds of atoms (elements). Scientists in labs have been able to artificially produce about 20 more. As their chemical properties have a periodic behaviour, elements are usually arranged in a tabular form, the so-called

"Periodic Table of the Elements", organized on the basis of their atomic number, electron configurations, and recurring chemical properties. The periodic table we use today is based on the one devised and published in 1869 by the Russian chemist Dmitri Mendeleev (Tobolsk, Siberia, 1834—Saint Petersburg, Russia, 1907), who found that he could arrange the 65 elements then known in a tabular form based on their atomic mass [3].

The standard form of the Periodic Table consists of a grid of elements laid out in 18 columns and 7 rows, with a double row of elements—lanthanides and actinides—below that (see Fig. 1.5). Elements are shown from top to bottom in order of increasing atomic number, which is listed with the chemical symbol in each box. The rows of the table are called periods; the columns are called groups, which have names such as alkali metals, alkali earth metals, transition metals, post-transition metals, nonmetals, halogens, metalloids, or noble gases. Elements of each group have the same number of electrons in the outermost shell, and consequently have similar chemical behaviour. For instance, the elements on the first column (Li, Na, K, Rb, Cs e Fr) behave like hydrogen (H) in chemical reactions as if they had one single electron, called the *valence electron*. Analogously, the elements on the last column (helium, neon, argon, krypton, xenon and radon) are all inert gases, because their atomic electron configuration is such that they have little tendency to

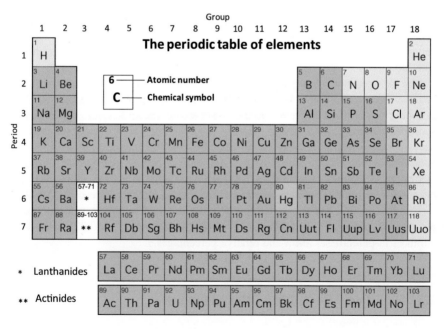

Fig. 1.5 Periodic table of the elements. The number in the top left corner of each box is the atomic number. Elements in green, pale blue and yellow boxes are solid, liquid and gas at room temperature, respectively. Adapted from [5]

participate in chemical reactions. Most elements in the Periodic Table are metals; they are malleable, shiny, and ductile. Nonmetal elements lack metallic properties and are volatile. Metalloids share properties with both nonmetals and metals (see Ref. [4] for more details).

1.4   Nuclear Size and Density

In first approximation, protons and neutrons in the nucleus are packed to form a roughly spherical region with radius R given approximately by:

$$R \approx r_0 A^{1/3}, \tag{1.2}$$

with $r_0 = 1.2 \times 10^{-15}$ m $= 1.2$ fm. The unit length equal to 1 femtometre ($=10^{-15}$ m) is called fermi and was so named in honour of the Italian physicist Enrico Fermi (Rome, Italy, 1901—Chicago, Illinois, USA, 1954. Nobel laureate in Physics in 1938). It is a typical length scale of nuclear physics. It is worth noticing that a nucleus has no clearly defined boundary, and hence the nuclear radius has only an indicative meaning.

Equation (1.2) indicates that the volume of the nucleus, V, is proportional to the mass number A, namely to the number of nucleons it contains:

$$V = \frac{4}{3}\pi R^3 = \text{using Eq. (1.2)} = \frac{4}{3}\pi r_0^3 A. \tag{1.3}$$

As one adds protons and neutrons in a nucleus, to form, respectively, new isotones or new isotopes, the nuclear volume simply grows with the number of nucleons in such a way that the density of nucleons (number of nucleons per unit volume) is constant. In other words, each proton and neutron occupies essentially the same volume ($V_{nucleon} = 7.2$ fm^3, see Problem 1.2), regardless of the number of nucleons in the nucleus. Therefore, the nucleons in a nucleus are packed together to form a mass of incompressible material. In this respect, nuclei are very different from atoms.

The mass number A is, in turn, directly proportional to the total nuclear mass, since—as said above—neutrons and protons have approximately the same mass. Therefore, the volume and mass of a nucleus are directly proportional, and consequently also the mass density of nuclear matter is approximately the same for all nuclei.

Thus, if nucleus A has twice the radius of nucleus B, then nucleus A will have eight times (2^3) as many nucleons as nucleus B. This is what happens with any material that cannot be compressed: if a solid sphere has twice the radius of another sphere made of the same material, it will have eight times the mass. The volume,

Table 1.2 Values of the density of some substances

Substance	Density [kg/m^3]
Nuclear matter	$\sim 10^{17}$
Centre of the Sun	$\sim 10^5$
Uranium	18.7×10^3
Mercury	13.59×10^3
Lead	11.3×10^3
Iron	7.86×10^3
Aluminium	2.70×10^3
Water (4 °C)	1000
Air	1.29
Laboratory ultra-high vacuum	$\sim 10^{-15}$
Interstellar space	$\sim 10^{-21}$
Intergalactic space	$\sim 10^{-27}$

and hence mass, of the sphere is proportional to the cube of its radius. The fact that nuclei obey this rule suggests that indeed all nuclei are made out of incompressible material of the same density.

The fact that nuclear densities do not increase when increasing the mass number A, in turn, implies that a nucleon does not interact with all other nucleons inside the nucleus, but only with its nearest neighbours. This phenomenon is an aspect of a very important property of the force between nucleons called the saturation of nuclear forces (see Sect. 1.5).

It is remarkable that atoms and their nuclei behave so differently. In contrast to nuclei, the size of an atom does not increase with the atomic number Z, implying that the electron density does increase with Z. This is due to the long-range Coulomb attraction of the nucleus for the electrons.

In Table 1.2 the values of the density of some substances are given for a useful comparison. The density of nuclear matter is about 10^{14} times (i.e. about hundred thousand billion times) denser than ordinary matter. If a solid football (a sphere of about 22 cm diameter) were made of pure nuclear matter, it would weigh as much as a few millions of the largest cruise ships.[4]

1.4.1 Insight: Measuring the Nuclear Radius

It is possible to determine the spatial distribution of nucleons inside a nucleus by scattering electrons off the nucleus. Since electrons have negative electric charge, a

---

[4]The world's largest cruise ship (the Royal Caribbean Harmony, presently under construction, scheduled for launch in April 2016) has a mass of about 2.28×10^8 kg. A nuclear matter football would have a mass of 5.6×10^{14} kg, i.e. 2.4×10^6 times that of the Royal Caribbean Harmony ship.

high-energy electron moving through a nucleus is deflected by the positively charged protons in a way that depends on how protons are distributed. This allows to reconstruct the charge density (or, equivalently, the proton probability distribution), $\rho(r)$. Assuming that protons are evenly distributed throughout the volume of the nucleus, this allows measuring the size of a nucleus.

Figure 1.6 shows the charge densities $\rho(r)$ of various nuclei as a function of the distance from the nuclear centre, as determined in elastic electron-nucleus scattering ("elastic" means that the scattering does not alter the internal configuration of the nucleus (see Sect. 3.1)). As illustrated in the figure, the charge distributions in the extremely light nuclei of hydrogen and helium are peaked at $r = 0$, in the light nuclei of carbon and oxygen they are peaked at about 1 fm, while in heavy nuclei they are flat up to 4–5 fm, then fall to zero over a distance of about 2 fm. Moreover, there is a large disparity between average central densities for the proton and all other nuclei. The helium nucleus is also a unique case as it exhibits a much larger central density than all heavier nuclei.

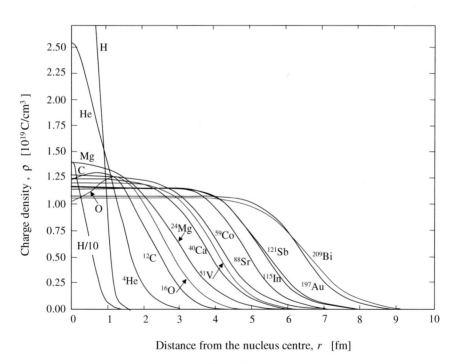

Fig. 1.6 Experimental charge density ρ [10^{19} C/cm^3] as a function of the distance from the centre of the nucleus, r [fm], as determined in elastic electron–nucleus scattering. All nuclei except the very lightest have about the same density in their interior. Light nuclei have charge distributions that are peaked at $r = 0$ or $r \sim 1$ fm, while heavy nuclei have flat distributions up to 4–5 fm, that fall to zero over a distance of about 2 fm. The *curves* in this figure are obtained from data of Refs. [6, 7]

Apart from the diffuse surface, all nuclei with $A > 40$ have about the same value of the charge density distribution (between 1.1×10^{19} and 1.3×10^{19} C/cm^3), independently of the nucleus under consideration. This corresponds to about 0.075 protons per fm^3. Assuming the neutron and the proton densities to be the same, one finds an almost identical nucleon density of $\rho_0 \approx 0.15$ nucleons/fm^3 in the nuclear interior for nearly all nuclei (see Problem 1.2).

From the above discussion, it is evident that nuclei are not spheres with a sharply defined surface. If one defines the nuclear radius as the distance from the centre of the nucleus to a point where the density is half that at the centre, one finds a value that corresponds to a good approximation to that of Eq. (1.2).

Problem 1.1: Nuclear size. Calculate the radius of the nuclei hydrogen (^{1}H$_1$), aluminium (^{27}Al$_{13}$), and uranium-238 (^{238}U$_{92}$).
Solution: Using Eq. (1.2) one gets:

$$R\left(^1\text{H}_1\right) = 1.2 \times 10^{-15}\text{m},$$

$$R\left(^{27}\text{Al}_{13}\right) = (1.2 \times 10^{-15}\text{m}) \cdot (27)^{1/3} = 3.6 \times 10^{-15}\text{m},$$

$$R\left(^{238}\text{U}_{92}\right) = (1.2 \times 10^{-15}\text{m}) \cdot (238)^{1/3} = 7.4 \times 10^{-15}\text{m}.$$

As nuclear radii vary as $A^{1/3}$, the largest nucleus in nature, ^{238}U, has a radius only six times greater than that of the smallest nucleus, ^{1}H.

Problem 1.2: Nuclear density. Calculate the density of nuclear matter D, and compare it with that of everyday objects.
Solution: In a nucleus the number of nucleons per unit volume ρ_0 is:

$$\rho_0 = \frac{A}{V} = \frac{A}{\frac{4}{3}\pi R^3} = \frac{3A}{4\pi r_0^3 A} = \frac{3}{4\pi r_0^3} = \frac{3}{4 \cdot 3.14 \cdot (1.2 \times 10^{-15})^3}$$
$$= 0.138 \times 10^{45} \text{ nucleons/m}^3.$$

This value (equal to 0.138 nucleons/fm^3) is slightly smaller than that (0.15 nucleons/fm^3) obtained from electron scattering off nuclei.
One gets the nuclear matter density by multiplying ρ_0 by the average mass of the nucleon:

$$D = (0.138 \times 10^{45}\text{nucleons/m}^3) \times (1.67 \times 10^{-27}\text{kg}) = 2.30 \times 10^{17}\text{kg/m}^3.$$

It is a huge value, equal to 230,000 tonnes per mm^3. Just as a comparison, the density of water at room temperature is 1000 kg/m^3, and the density of lead is 11,300 kg/m^3.

1.5   The Nuclear Force and the Diagram of Nuclei

The number of protons and neutrons that can give rise to stable nuclei is not arbitrary. Some combinations yield nuclei that last forever, others are unstable and are subject to radioactive decay (see Chap. 2).

This is shown in Fig. 1.7 that is a diagram of the number of protons Z as a function of the number of neutrons N for the currently known nuclei. These are about 3200 [8] and are largely produced naturally in the stars and artificially in nuclear reactors and in the laboratory. The number of nuclei that have been created and studied increases year by year. It is thought that some 6000 combinations of protons and neutrons can exist, albeit fleetingly in many cases.

In Fig. 1.7, the black squares denote the stable nuclei and also nuclei with lifetime considerably longer than the age of the solar system. The curve passing through these points is called *the stability curve*. The coloured squares indicate the unstable nuclei. The black line represents the bisector of the axes, and is described

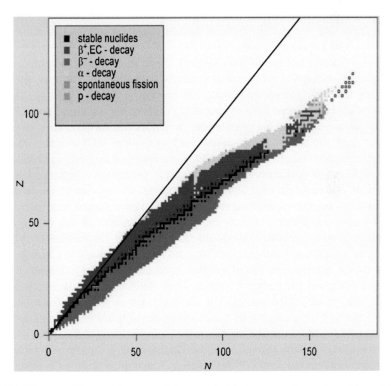

Fig. 1.7 Diagram of the nuclides presently known: the black squares denote the stable nuclei and nuclei with lifetime considerably longer than the age of the solar system; the coloured squares denote the unstable (i.e. radioactive) nuclei. All stable nuclei (and also almost all unstable nuclei) lie below the straight line $N = Z$ (black line), which reflects the fact that the number of neutrons becomes greater than the number of protons as the atomic number Z increases [10]

by the equation $Z = N$. The isotopes of the same element are on the same horizontal line, the isotones on the same vertical line.

This sort of diagram is called a "Segrè chart", in honour of the Italian-American physicist Emilio G. Segrè (Tivoli, Italy, 1905—Lafayette, California, USA, 1989) who expanded a similar representation first published by Fea in 1935 [9].

As it is seen in the Fig. 1.7, most nuclei are unstable or, as it is usually said, *radioactive*. Generally, for each mass number $A = (N + Z)$, there are only one or two nuclides (i.e. combinations of (N, Z) values) sufficiently long-lived to be naturally present on Earth in significant quantities.

The figure also shows that the light nuclei ($A < 20$) generally contain almost an equal number of protons and neutrons, while for heavier nuclides the number of neutrons N is always greater than that of the protons Z, and the neutron excess tends to grow when Z increases. This is because the neutrons act as a kind of glue that keeps repelling protons together. The greater the repelling charge, the more glue is necessary. Therefore, slightly more than one neutron is required per proton and as the mass number A increases so does the neutron-proton ratio. Indeed, carbon-12 has six protons and six neutrons; iron-56 has 26 protons and 30 neutrons; uranium-238 has 92 protons and 146 neutrons. In other words, the stable nuclei lie along a curve (the one passing through the black squares in Fig. 1.7) starting out with $N/Z = 1$ and ending up with $N/Z = 1.6$.

The blue squares in Fig. 1.7 indicate nuclei with an excess of neutrons, which, consequently, undergo a process that converts a neutron to a proton, therefore decreasing the neutron-to-proton ratio: this process is called β^--decay. The red squares indicate nuclei that have an excess of protons; they undergo a so-called β^+-decay, which converts a proton to a neutron, thereby increasing the neutron-to-proton ratio; they can also undergo a so-called Electron Capture (EC) process, where an atomic electron is absorbed by the nucleus, again converting a proton to a neutron. The yellow nuclei are those that decay by emitting alpha particles (i.e. helium nuclei with two protons and two neutrons) and the green nuclei undergo spontaneous fission. There are a few orange nuclei along the upper edge of the coloured area: these nuclei decay by emitting protons. The open, uncoloured squares at the top right are super-heavy nuclei recently produced in laboratory experiments. All the above-mentioned decay processes are discussed more extensively in Chap. 2.

Protons and neutrons are bound together in the nucleus by a fundamental force that has been given the name of *nuclear force* because it operates between the particles that make up the nuclei. The nuclear force is different from both the electric force and the gravitational force: at distances typical of neutrons and protons in nuclei ($\sim 10^{-15}$ m), the nuclear force is much stronger than both (for this reason it is also called the *strong force*), but decreases more rapidly with increasing the distance between nucleons. If the nucleons move away from each other more than few fermis, the nuclear attraction practically disappears. This fact is usually expressed by saying that the nuclear force has a short-range action. The action radius, or range, of the nuclear forces is $r_a \cong 1.5$ fm. Moreover, the nuclear force is charge independent. This means that the nuclear forces acting between two protons, or between two neutrons, or between a proton and a neutron, are the same.

The short range of the nuclear force plays an important role in the stability of the nucleus. This stability is based on a balance of forces. In fact, for a nucleus to be stable, the electrostatic repulsion between the protons must be balanced by the attraction between the nucleons due to the nuclear force. However, while every proton exerts a repulsive electrical force on every other proton in the nucleus since the electrostatic force has a very long range of action, every nucleon (proton or neutron) attracts only the nucleons that are at a distance less than the very-short range r_a of the nuclear force. Therefore, when the number Z of protons in the nucleus increases, the number N of neutrons must grow even more rapidly so that the stability is maintained.

However, a point is reached where it is no longer possible to find a balance between the attractive nuclear force and the repulsive electrostatic force by increasing the number of neutrons, and the extra neutrons can no longer provide extra nuclear force to balance the long-range electrostatic repulsion of the protons. Therefore, in this extreme situation the nucleus becomes unstable.

As already briefly discussed above, for elements up to calcium, stable isotopes have equal or similar numbers of protons and neutrons. For heavier elements, stable nuclei contain more neutrons than protons and the excess of neutrons increases with atomic number. For example, stable isotopes of lead, which has 82 protons, contain 122, 124, 125 or 126 neutrons. The stable nucleus with the maximum number of protons is the isotope bismuth, $^{209}\mathrm{Bi}_{83}$, which has 83 protons and 126 neutrons.[5] The nuclei with more than 83 protons, like uranium which has Z = 92, are unstable, i.e. they disintegrate spontaneously or reorder their internal structure with time. This phenomenon is called *radioactivity*.

1.6   Nuclear Masses and Mass Defect

A nucleus with Z protons and A nucleons has a mass $M(Z, A)$ slightly smaller than the sum of the masses of the A nucleons that it contains:

$$M(Z,A) < ZM_p + NM_n, \tag{1.4}$$

where M_p and M_n are the mass of the proton and the neutron, respectively. The missing mass

$$\Delta m = ZM_p + NM_n - M(Z,A), \tag{1.5}$$

historically called the *mass defect*, is of fundamental importance in the study of strongly interacting bound systems. It indicates the degree of binding of protons and neutrons in the nucleus.

---

[5]Actually, it was recently discovered that ^{209}Bi undergoes decay but with an extremely long lifetime of the order of 10^{19} years, so it is defined as "virtually stable".

This fact is quite general: the mass of any bound system M is less than the sum of the masses of its constituents when they are separated, because one has to provide energy to separate them. For instance, the hydrogen atom has a mass which is 13.6 eV/c^2 (see Sect. 1.7 for a definition of this mass unit) less than the sum of the proton and electron masses, which implies that 13.6 eV must be supplied to ionise hydrogen, i.e. to separate it into unbound proton and electron. In most cases, the missing mass Δm is so small that cannot be determined by direct measurements of the mass and one must therefore resort to other methods. For example, for the system Earth-Sun, $\Delta m/M \approx 10^{-17}$; for a crystal, $\Delta m/M \approx 10^{-11}$; for atomic hydrogen in the *ground state* (the quantum configuration with the lowest energy), $\Delta m/M \approx 1.5 \times 10^{-8}$. In the most violent chemical reactions (*e.g.* in explosions) the energies released do not exceed the 10^{-8} fraction of the mass of the reactants. The order of magnitude of the nuclear binding is vastly larger than the other bounds ($\Delta m/M \approx 1 \%$), but its role in determining the total mass of the system is the same.

For example, for the deuteron, with mass $M_d = 3.343584 \times 10^{-27}$ kg [2], one has

$$\Delta m = M_p + M_n - M_d = 3.966 \times 10^{-30}\,\text{kg}.$$

The difference is rather small, but noticeable on the nuclear scale: $\Delta m \approx 0.0024 \times M_p = 0.0012 \times M_d$.

It is convenient to include the mass of the Z atomic electrons into the right-hand side of Eq. (1.5), and use the atomic masses since they can be measured to a considerably higher precision than nuclear masses. Then the Eq. (1.5) becomes:

$$\Delta m = Z M_\text{H} + N M_n - M_\text{A}(Z, A). \tag{1.6}$$

Here $M_\text{H} = (M_p + m_e)$ is the mass of the hydrogen atom, m_e is the mass of the electron, and $M_\text{A}(A,Z)$ is the mass of the atom with Z electrons whose nucleus contains A nucleons. We notice that using atomic masses the mass of Z electrons is included $[M_\text{A}(A,Z) = M(A,Z) + Z m_e;\ M_\text{H} = M_p + m_e]$. In the calculations they cancel out with the masses of the Z electrons of the Z hydrogen atoms. The mass change of the Z electrons in the atom because of the binding energy in the Z element is below the determination error of the masses, and is therefore negligible.

Table 1.3 shows the masses of neutral atoms (including electrons and taking into account their binding energy) of some isotopes. These are usually given in atomic mass units (often abbreviated amu, symbol u), equal to 1/12 of the mass of an atom of $^{12}\text{C}_6$ (1u $= 1.66054 \times 10^{-27}$ kg [2]).

Since mass and energy are equivalent, in nuclear physics it is customary to measure masses of particles in the units of energy; namely, in eV/c^2 (see next section for the explanation of the c^2 factor). Recalling that the electronvolt is the kinetic energy that an electron acquires when it is accelerated by a potential difference of 1 V (1 eV $= 1.602 \times 10^{-19}$ J), it is easy to derive that 1 u $= 931.494 \times 10^6$ eV/c^2 (see Problem 1.3). Widely used are the multiple of the

Table 1.3 Mass of neutral	Nucleus	Symbol	Mass [u]
atoms (including electrons and taking into account their binding energy) of some isotopes [11]	Hydrogen	$^1H_1$	1.007825
	Deuterium	$^2H_1$	2.014102
	Tritium	$^3H_1$	3.016049
	Helium	$^4He_2$	4.002603
	Carbon	$^{12}C_6$	12.000000
	Nitrogen	$^{14}N_7$	14.003074
	Oxygen	$^{16}O_8$	15.994915
	Aluminium	$^{27}Al_{13}$	26.981538
	Calcium	$^{40}Ca_{20}$	39.962591
	Iron	$^{56}Fe_{26}$	55.934939
	Copper	$^{63}Cu_{29}$	62.939598
	Silver	$^{107}Ag_{47}$	106.905092
	Lead	$^{206}Pb_{82}$	205.974440
	Uranium	$^{238}U_{92}$	238.050784

electronvolt: the kiloelectronvolt (keV), equal to 1000 eV; the megaelectronvolt (MeV), equal to 10^6 eV; and the gigaelectronvolt (GeV), equal to 10^9 eV.

The standard unit of energy, the joule, is too large to measure the energies associated with individual nuclei. This is why in nuclear physics it is more convenient to use the MeV, which is a much smaller unit (1 MeV = 1.602×10^{-13} J). Thus, in the unit of MeV most of the energies in nuclear world are expressed by values with only a few digits before decimal point and without powers of tens.

It is important to notice that very often in particle and nuclear physics one uses a unit system in which $c = 1$ so both mass and energy are measured in eV. We will adopt this unit system also throughout this book.

1.7   The Mass-Energy Equivalence

According to relativity, the energy of a body at rest, E_0, is related to its mass M by the relationship

$$E_0 = Mc^2, \tag{1.7}$$

where $c = 2.99792 \times 10^8$ m/s is the speed of light in vacuum. In the following, in calculations we will often use, for convenience, the approximate value $c \approx 3 \times 10^8$ m/s. E_0 is called the rest energy of the body.

It is customary to refer to Eq. (1.7) as *the equivalence of mass and energy*, or simply *mass-energy equivalence*, because one can choose units in which $c = 1$, and hence $E_0 = M$.

It is worth noting that the rest energy of familiar objects is tremendous. A 0.055 kg tennis ball has a rest energy of $(0.055 \text{ kg}) \times (3 \times 10^8 \text{ m/s})^2 = 4.95 \times 10^{15}$ J, which is more than ten thousand billion times as much as its kinetic energy at a speed of 70 m/s, $T = \frac{1}{2}(0.055 \text{ kg}) \times (70 \text{ m/s})^2 = 134.75$ J. An electron has a rest energy of 511×10^3 eV, which is over ten times the kinetic energy it attains in a 50 kV X-ray tube.

Equation (1.7) says that the mass can be converted into energy and vice versa. All the development of physics and chemistry preceding the theory of relativity was based on the assumption that the mass and energy of a closed system are conserved in all possible processes and that they are separately conserved (i.e. mass is conserved and energy is conserved). According to relativity, instead of having a law of conservation of rest mass and another law of conservation of energy we only have one law, the conservation of energy, in which the energy includes the rest energy (i.e. the mass).

As previously mentioned, in chemical reactions the fraction of the mass that transforms into other forms of energy (and vice versa), is so small that it is not detectable by a direct measurement of mass, even in very precise measurements (see Problem 1.5). In nuclear processes, however, the energy release is very often million times higher and hence observable.

Problem 1.3: Conversion factor of energy units. Calculate the conversion factor between the mass atomic unit and the electron-volt/c^2 (eV/c^2).

Solution: According to its definition $1 \text{ u} = \frac{1}{12} M(^{12}\text{C}) = 1.66054 \times 10^{-27}$ kg. Using the mass-energy relation $E_0 = mc^2$, with $c = 2.99792 \times 10^8$ m/s, the energy associated to this mass is:

$$E_0 = mc^2 = (1.66054 \times 10^{-27} \text{ kg}) \times (2.99792 \times 10^8 \text{ m/s})^2$$
$$= 1.49241 \times 10^{-10} \text{ J} .$$

Remembering that one electronvolt is the amount of energy that an electron gains when it moves across an electric potential difference of one volt $(1 \text{ eV} = 1.60218 \times 10^{-19} \text{ C} \times 1 \text{ V} = 1.60218 \times 10^{-19} \text{ J})$, it follows

$$E_0 = \frac{1.49241 \times 10^{-10} \text{ J}}{1.60218 \times 10^{-19} \text{ J/eV}} = 931.487 \times 10^6 \text{ eV} = 931.487 \text{ MeV}.$$

Then

$$1 \text{ u} = 931.487 \text{ MeV}/c^2.$$

i.e. one atomic mass unit is the amount of mass that, when multiplied by c^2, gives an energy of 931.487 MeV. (By using all digits in the above quantities, one can find a value of 931.494 MeV/c^2, which is the one reported in [2]).

Problem 1.4: Proton-proton fusion. The "proton-proton chain reaction" is the chief source of the energy radiated by the Sun. It starts with protons and, through a series of steps, turns them into helium nuclei. The first step of the chain is the reaction $p + p \rightarrow d + e^+ + \nu$, where two protons fuse forming a deuteron and emitting one positron (e^+) and one neutrino (ν). (The positron is the antiparticle of the electron: it is positively charged and has the same mass and magnitude of the charge as the electron. The neutrino appears in the so-called nuclear β-decay. For our purpose, the neutrino has no mass; we say it is *massless*.) Calculate the total energy E released in this thermonuclear fusion reaction (mass of deuteron 1875.61 MeV).

Solution: With clear meaning of the symbols, the total energy released, E, is

$$
\begin{aligned}
E &= \left(M_p + M_p - M_d - m_{e+}\right)c^2 \\
&= 2 \times 938.272\,\text{MeV} - 1875.61\,\text{MeV} - 0.510999\,\text{MeV} = 0.42\,\text{MeV}.
\end{aligned}
$$

The deuteron, the positron and the neutrino flying away share this energy.

Problem 1.5: Mass defect in a macroscopic process. Two lumps of plasticine, each with mass $m = 1.0$ kg and speed $v = 2.5$ m/s, are moving in opposite directions. They collide and stick together forming a stationary blob. In the collision, the kinetic energy of the two lumps disappears, so their rest mass must increase to some value M. Calculate the size of the mass increase of the blob, assuming that all the kinetic energy of the two lumps transforms into rest mass of the blob.

Solution: We can evaluate the final mass M of the two lumps together using the law of conservation of energy: the total energies before and after the collision must be the same

$$
(mc^2 + \frac{1}{2}mv^2) + (mc^2 + \frac{1}{2}mv^2) = Mc^2 + 0 .
$$

This can be rearranged to give the change in the rest mass

$$
M - 2m = \frac{mv^2}{c^2} = (1.0\,\text{kg})(\frac{2.5\,\text{m/s}}{3.0 \times 10^8\,\text{m/s}})^2 = 6.9 \times 10^{-17}\,\text{kg}.
$$

The change in the rest mass is very small. In reality, in this example from the macroscopic world, the kinetic energy will be mainly transformed into acoustic (sound) and thermal energy. However, in nuclear physics, two nuclei can collide and form a nucleus with mass higher than the sum of the masses of the two initial nuclei. This happens if the two initial nuclei have sufficient kinetic energy, all or part of which is indeed transformed into rest mass of the new system.

1.8   Nuclear Binding Energy

The nuclear binding energy $B(Z, A)$ of a nucleus with Z protons and $N = (A\text{-}Z)$ neutrons is defined as the difference between the masses of the nuclear constituents and the nuclear mass:

$$B(Z,A) \ = \ \Delta mc^2 \ = \ \left[ZM_p + NM_n - M(Z,A) \right] c^2, \tag{1.8}$$

or, using the atomic masses:

$$B(Z,A) \ = \ \left[ZM_H + NM_n - M_A(Z,A) \right] c^2. \tag{1.9}$$

Often the binding energy is defined as the difference between the nuclear mass and the masses of the nuclear constituents,

$$B'(Z,A) = \left[M_A(Z,A) - ZM_H - NM_n \right] c^2. \tag{1.8'}$$

Clearly $B'(Z,A) = -B(Z,A)$ and is always negative.

$B(Z, A)$ represents the work necessary to disassemble the nucleus into separated nucleons or, conversely, the energy that would be liberated when Z protons and N neutrons combine to form a nucleus. If the mass of a nucleus were exactly equal to the sum of the masses of the constituent Z protons and N neutrons, the nucleus could disintegrate without the need for any energy supply. For a nucleus to be stable, its mass must be less than that of its constituents so that one should provide energy to split it.

In order to understand which nuclear isotopes are most tightly bound and to compare light and heavy nuclei on an equal basis, it is useful to examine a plot of the binding energy per nucleon B/A versus the mass number A. A tightly bound nucleus is one in which each nucleon in the nucleus has a large binding energy, regardless of how many nucleons there are in the nucleus.

Figure 1.8 shows the trend of the average binding energy per nucleon B/A as a function of the mass number A, for $1 \leq A \leq 238$. Each point in the plot represents a stable or nearly stable (i.e. with lifetime considerably longer than the age of the solar system) isotope.

Observing the figure it is noted that, starting from the lighter elements, the binding energy per nucleon rapidly increases with A, reaching the maximum value of about 8.8 MeV per nucleon for the isotope $^{62}Ni_{28}$ ($B/A(\text{Ni}) = 8.7946 \pm 0.0003$ MeV/nucleon), and then it decreases monotonically down to 7.3 MeV/nucleon for ^{238}U. This small value—as we will see shortly—is insufficient to keep the nucleus stable. This trend is of fundamental importance for nuclear decays and for the production of energy from nuclear processes. Very light nuclei, when fused with each other, form more tightly bound systems, as do very heavy nuclei when they split up into lighter fragments.

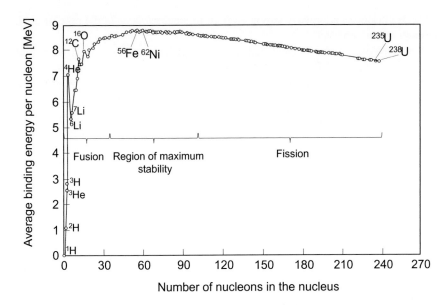

Fig. 1.8 Experimental values of the average binding energy per nucleon as a function of the mass number A. The nuclei in the region of ^{56}Fe and ^{62}Ni have the highest values of binding energy per nucleon

The trend of the average binding energy per nucleon can be easily understood by recalling the features of the attractive nuclear force between nucleons and of the repelling electrostatic force between protons. As said in Sect. 1.5, the attractive force on a single nucleon is due to all the other nucleons of the nucleus that are within the effective range of nuclear force: the higher this number the greater the attractive force and thus the energy required to tear off a nucleon. This explains the lower value of the binding energy of light nuclei which have a small number of nucleons. They are small enough that most of the nucleons feel the attractive nuclear force from all the others, so that adding more nucleons tends to produce a tighter binding. This explains the increase of B/A with A for light nuclei ($A < 30$).

On the contrary, in a very large nucleus every proton exerts a repulsive electrical force on every other proton, while every nucleon does not attract every other nucleon, because some of them are too far away for the nuclear force to act between them. In other words, in a large nucleus, the nucleons on one side of the nucleus are too far away to be attracted by those on the opposite side. The result is a weaker binding for very large nuclei. This effect is seen on the graph as the gradual decrease of binding energy per nucleon above $A = 60$.

Figure 1.8 also shows that some nuclei are exceptionally strongly bound compared to nuclei of similar A. It is the case for ^{4}He, ^{12}C, ^{16}O. This is because these nuclei have certain quantum levels completely filled, which makes attraction between nucleons particularly strong (similar to what happens in the case of noble gases in atomic physics).

The nucleons in the nucleus helium-4 ($^4\text{He}_2$) have indeed a relatively high binding energy, $B/A(^4\text{He}) = 7.06$ MeV (see Problem 1.7). Therefore, emission of ^{4}He nuclei—called α-particles in nuclear decays (see Sect. 2.1)—is energetically possible in heavy nuclei. This phenomenon is easy to understand in terms of binding energies. In a nucleus with $B/A = 8$ MeV (nuclides with mass number $A \approx 185$ in Fig. 1.8), a cluster of two protons and two neutrons has a total binding energy larger than that of a ^{4}He nucleus; namely, (4·8 MeV/nucleon) = 32 MeV, with respect to (4·7.06 MeV/nucleon) = 28.2 MeV. Therefore, only less than 4 MeV ($32-28.2 = 3.8$ MeV) are required to emit an α-particle from that nucleus, compared to 8 MeV required to emit a single nucleon. As in nuclei with mass number $A > 185$ the nucleon binding energy is $B/A < 8$ MeV and decreases with increasing A, the energy required for the emission of an α-particle can reduce to zero and the nucleus can emit it spontaneously, i.e. without the need for additional energy (see Problem 1.9). Clearly, for two protons and two neutrons it is energetically advantageous to be bound in a nucleus ^{4}He rather than in a much heavier nucleus, where the two protons would suffer the repulsive force of all other protons. This is the reason why nuclei heavier than $^{209}\text{Bi}_{83}$ are unstable and emit α-particles spontaneously.[6]

The nucleons in the ^{12}C and ^{16}O are even more strongly bound (7.6 and 8.0 MeV per nucleon, respectively) than are in helium and as a result decay by ^{12}C or ^{16}O emission is also energetically possible in some heavy nuclei.

The curve of Fig. 1.8 also shows that the fusion reaction—that is, the reaction in which two light nuclei combine to form a single heavier nucleus—is energetically favourable for light nuclides (see Problems 1.4, 1.11 and 1.12), while fission—that is the splitting of a heavy nucleus into two or more smaller nuclei—is for heavy nuclei. A nucleus is, in fact, dynamically unstable when its binding energy per nucleon is lower than that in the fragments in which it can divide. An examination of Fig. 1.8 shows that this situation occurs for all nuclei of mass number $A > 100$, which are, therefore, unstable with respect to the fission process (see Sect. 3.3).

Figure 1.9 illustrates in a pictorial way the mass unbalance between the initial and final state in a fusion and a fission process, respectively. The balance on the left of the figure shows that in the fusion reaction $^2\text{H}_1 + {}^3\text{H}_1 = {}^4\text{He}_2 + n$, the sum of the masses of two light nuclei (deuterium, $^2\text{H}_1$, and tritium, $^3\text{H}_1$) is greater than the sum of the mass of the helium nucleus that forms in their fusion and that of the free neutron that is released. Such fusion reaction, therefore, converts the missing mass into energy. The balance on the right of the figure shows similarly that for the fission reaction $^{236}\text{U}_{92} \rightarrow {}^{95}\text{Sr}_{38} + {}^{139}\text{Xe}_{54} + 2n$, the sum of the masses of the two

---

[6]Even when the emission of an α cluster is energetically possible, i.e. it can happen spontaneously and is accompanied by the release of kinetic energy, the probability for its occurrence follows specific rules dictated by quantum mechanics. This is generally true for all energetically allowed nuclear processes.

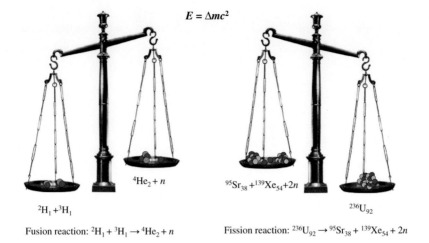

$$E = \Delta m c^2$$

Fusion reaction: $^2H_1 + {}^3H_1 \rightarrow {}^4He_2 + n$ Fission reaction: $^{236}U_{92} \rightarrow {}^{95}Sr_{38} + {}^{139}Xe_{54} + 2n$

Fig. 1.9 Pictorial illustration of the mass unbalance between the initial and final state in a fusion (*left*) and a fission (*right*) reaction

medium A nuclei (strontium, $^{95}Sr_{38}$, and xenon, $^{139}Xe_{54}$) and the two neutrons released is less than that of the heavy nucleus (uranium, $^{236}U_{92}$) which has generated them. In this case, the fission reaction converts the missing mass into energy.

Problem 1.6: Binding energy of the ^{4}He isotope. Calculate the mass defect and the binding energy of the most abundant isotope of helium, the nuclide $^4He_2$, whose mass is 6.6447×10^{-27} kg.
Solution: To get the mass defect Δm, one has to subtract the mass of the nucleus ^{4}He from the sum of the masses of two protons and two neutrons that make it up:

$$\Delta m\left(^4He\right) = 2M_p + 2M_n - M_e$$
$$= 2 \times \left(1.67262 \times 10^{-27}\,kg\right) + 2 \times \left(1.67493 \times 10^{-27}\,kg\right)$$
$$-6.6447 \times 10^{-27}\,kg = 0.0504 \times 10^{-27}\,kg.$$

This is equal to about 8 ‰ of the mass of helium, M_{He}:

$$\frac{\Delta m}{M_{He}} = \frac{0.0504 \times 10^{-27}}{6.6447 \times 10^{-27}} = 0.0076.$$

It is interesting to compare this value with the yield of a process of nuclear fission [$\Delta m/m \approx 0.1\,\%$] and with the yield of electromagnetic processes [$\Delta m/m \approx 10^{-8}$]. Thermonuclear fusion is clearly the most efficient process existing in nature for transforming mass into energy.

The binding energy of the nucleus ^{4}He is then (using the approximate value $c \approx 3 \cdot 10^8$ m/s for the speed of light in vacuum):

$$B\left(^4\text{He}\right) = \Delta mc^2$$
$$= \left(0.0504 \times 10^{-27}\,\text{kg}\right) \times \left(3.00 \times 10^8\,\text{m/s}\right)^2 = 4.53 \times 10^{-12}\,\text{J}.$$

Usually the binding energies are expressed in electron volts instead of joules (1 eV = 1.60×10^{-19} J). Then:

$$B\left(^4\text{He}\right) = 4.53 \times 10^{-12}/1.60 \times 10^{-19} = 28.3\,\text{MeV}.$$

This value is more than 2 million times the energy required to remove an electron from an atom of hydrogen, which is 13.6 eV.

The average binding energy per nucleon in a nucleus of helium-4 is then

$$\frac{B\left(^4\text{He}\right)}{A} = \frac{28.3\,\text{MeV}}{4\,\text{nucleons}} = 7.08\,\text{MeV}.$$

Problem 1.7: Binding energy of iron. Calculate the total binding energy of the most abundant isotope of iron, the nuclide $^{56}\text{Fe}_{26}$.

Solution: The total binding energy is obtained by subtracting the mass of the nucleus $^{56}\text{Fe}_{26}$ from the sum of the masses of 26 protons and 30 neutrons that make it up.

Using the values of Tables 1.1 and 1.3, and Eq. (1.6) (remember that, when using the values of the atomic masses to determine the mass defect of a nucleus, one must take into account the mass of the electrons. For this, one shall use the mass of the hydrogen atoms instead of that of the proton), we have:

$$\Delta m\left(^{56}\text{Fe}_{26}\right) = 26 M_\text{H} + 30 M_\text{n} - M_\text{Fe}$$
$$= 26 \times (1.007825\,\text{u}) + 30 \times (1.008665\,\text{u}) - (55.934939\,\text{u}) = 0.528461\,\text{u}.$$

The binding energy of the nucleus $^{56}\text{Fe}_{26}$ is then:

$$B\left(^{56}\text{Fe}_{26}\right) = \Delta m\left(^{56}\text{Fe}_{26}\right) c^2$$
$$= (0.528461\,\text{u})\,c^2 \times \left(931.494\,\left(\text{MeV}/c^2\right)/\text{u}\right) = 492.258\,\text{MeV}.$$

The average binding energy per nucleon in a nucleus of iron-56 is

$$\frac{B\left(^{56}\text{Fe}\right)}{A} = \frac{492.258\,\text{MeV}}{56\,\text{nucleons}} = 8.7903\,\text{MeV}.$$

Problem 1.8: Binding energy of the last nucleon. Calculate the binding energy b of the last neutron in the nucleus $^{13}C_6$ (mass 13.003355 u).
Solution: With a clear meaning of the symbols, we have:

$$\Delta m = M\left(^{12}C_6\right) + M_n - M\left(^{13}C_6\right)$$
$$= 12.000000\,u + 1.008665\,u - 13.003355\,u = 0.005310\,u.$$

Therefore, the binding energy of the last neutron is

$$b = \Delta mc^2 = (0.005310\ u)c^2 \times \left(931.494\left(MeV/c^2\right)/u\right) = 4.9462\ MeV.$$

It is interesting to note that this neutron is less bound than the other nucleons of the nucleus ^{13}C, which have an average binding energy of about 7.5 MeV (see Fig. 1.8).

Problem 1.9: Spontaneous α–decay of radium. Examine whether the radium nucleus, $^{226}Ra_{88}$, (atomic mass $M_{Ra} = 226.025410$ u), can emit spontaneously an α-particle thereby transforming in a radon nucleus, $^{222}Rn_{86}$, (atomic mass $M_{Rn} = 222.017578$ u).
Solution: The mass of the nucleus $^{226}Ra_8$ is greater than the sum of the masses of the nuclei $^4He_2$ and $^{222}Rn_{86}$

$$M_{Ra} - (M_{Rn} + M_{He}) = 226.025410\ u - (222.017578 + 4.002603)u$$
$$= 0.005229\ u.$$

The spontaneous disintegration of radium-226 is possible because its mass is greater than that of the disintegration products. The process is said "energetically possible".

Problem 1.10: Binding energy of the last 20 nucleons. For the nuclei with mass number $A = 180$ and $A = 200$ the average binding energy per nucleon is respectively of 8.0 MeV and 7.85 MeV. Calculate the average binding energy b of the 20 additional nucleons in the second nucleus.
Solution: Using the data, it results that:

$$180 \times 8.0 + 20b = 200 \times 7.85,$$

from which it follows:

$$b = 6.5\ MeV.$$

If two protons and two neutrons of these 20 nucleons formed a nucleus helium-4, their binding energy would increase to 7.08 MeV. Then, the emission of an α-particle is energetically favoured, because it would lead to an energy reduction of $4 \times (7.08 - 6.5) = 2.32$ MeV.

Problem 1.11: Energy to break up carbon. Calculate the energy ε that must be supplied to a nucleus ^{12}C in order to break it up into three α particles. *Solution*: To evaluate the energy ε, we must subtract the mass of the nucleus ^{12}C from the sum of the masses of three ^{4}He-nuclei. Using the values shown in Table 1.3, we get:

$$\begin{aligned} \varepsilon &= \left[3M\left(^4\text{He}\right) - M\left(^{12}\text{C}\right) \right] c^2 = (3 \times 4.002603 \text{ u} - 12 \text{ u})c^2 \\ &= (12.007809 \text{ u} - 12 \text{ u})c^2 = (0.007809 \text{ u}) \times (931.494 \text{ MeV/u}) \\ &= 7.274 \text{ MeV}. \end{aligned}$$

Instead, when three α particles combine to form a ^{12}C nucleus, 7.274 MeV are released. This process is particularly important for the formation of carbon nuclei in stars.

Problem 1.12: Fusion energy release. Calculate how much energy is released in the fusion reaction $^2\text{H}_1 + {}^3\text{H}_1 \rightarrow {}^4\text{He}_2 + n$, where a nucleus of deuterium and one of tritium fuse forming a helium nucleus and emitting a free neutron.
Solution: The energy released is obtained by subtracting the total mass of the final system (nucleus $^4\text{He}_2$ and a neutron) from the total mass of the initial system (deuterium and tritium):

$$\begin{aligned} &M\left(^2\text{H}\right) + M\left(^3\text{H}\right) - M\left(^4\text{He}_2\right) - M_\text{n} \\ &= (2.014102 + 3.016049 - 4.002603 - 1.008665)\text{u} = 0.018883 \text{ u}. \end{aligned}$$

Therefore, the energy released is $(0.018863 \text{ u}) \cdot (931.494 \text{ MeV/u}) = 17.59$ MeV.
As five nucleons participate to this fusion reaction, the energy released per nucleon is:

$$\frac{17.59}{5} = 3.52 \text{ MeV per nucleon}.$$

This energy is about 4 times greater than that released in a fission process, which is equal to about 0.9 MeV/nucleon (see Sect. 3.4.1). With the same amount of fuel, fusion reactions release more energy than fission ones.

1.9 The Nuclear Valley of Stability

The unstable nuclei, those indicated with red or blue squares in the Fig. 1.7, have a proportion of protons and neutrons that does not allow them to be energetically stable. As a consequence they undergo a transition to a stable configuration by transmuting to another nucleus, changing protons to neutrons, or neutrons to protons, while keeping the same atomic mass number. This process, called β-decay, will be discussed in more detail in Sect. 2.1.

The wrong balance of protons and neutrons gives the nucleus too much energy. This can be observed by examining the detailed behaviour of the atomic mass for isobar nuclei. As an example, Fig. 1.10 shows the nuclear masses of the isobars with mass number $A = 101$ versus the number of protons Z: the points correspond to different nuclides and the smooth parabolic curve connecting the points indicates the dominant behaviour of the nuclear binding energy as a function of the atomic number Z at fixed mass number A. The ruthenium isotope $^{101}Ru_{44}$, which has the lowest value of the atomic mass, is the only stable isobar with $A = 101$. The nuclei up the sides of the parabola are unstable; they eventually lose energy, falling down to the bottom of the parabola in a series of beta decays, transforming into the ruthenium isotope. The nuclides with higher Z than Ru decay by β^+ decay (thereby decreasing Z by one unit at each decay), nuclides with lower Z decay by β^- decay (thereby increasing Z by one unit at each decay).

This general pattern is followed throughout the periodic table. If we repeat the above procedure for different values of A, and plot a diagram of nuclides in a three-dimensional graph, in which the vertical energy scale corresponds to the

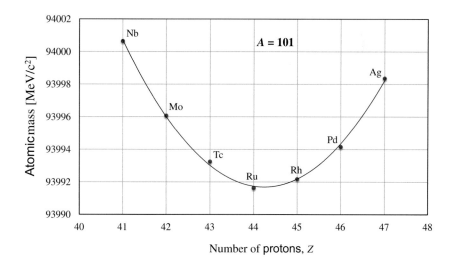

Fig. 1.10 The atomic mass of the isobars of $A = 101$ as a function of the atomic number Z in the region of the curve of stability. The nucleus ruthenium (Ru) is stable; all the others (niobium, molybdenum, technetium, rhodium, palladium and silver) are radioactive

Fig. 1.11 Rest mass energy dependence of nuclei on the numbers of proton, Z, and neutrons, N. All stable nuclei lie along the bottom of a valley. The bottom follows the shape of the curve $-B/A$ of the Fig. 1.8 and the sides of the valley curve upwards with a parabolic form as shown for mass number $A = 101$ in Fig. 1.10 [12]

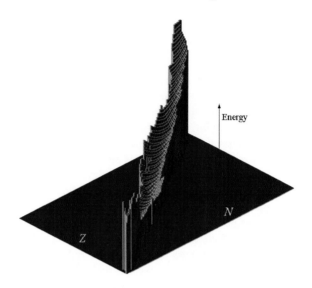

rest-mass energy of the nucleus (atomic mass), the mass parabolas for different A form a valley in the Z-N plane (see Fig. 1.11). All stable nuclei lie along the bottom of the valley, which is called the *valley of stability*.

The bottom of the valley follows the behaviour of the average binding energy per nucleon with reversed sign (B/A curve of Fig. 1.8 with a minus sign in front of it). It is not level, but slopes toward a lowest point that is near nuclei in the region of iron and nickel, which have less energy per nucleon (i.e. more binding energy per nucleon) than any other nucleus. The bottom of the valley also slopes upward slowly in the direction of the heaviest nuclei, such as uranium, and more steeply towards the lightest nuclei.

If a nucleus is off the bottom of the valley of stability, it rolls towards the bottom of the valley by changing protons to neutrons (or neutrons to protons), while keeping A constant. This process of losing excess energy is rather like falling down a hill and this diagram shows how the energy increases rapidly away from the stable nucleus at the bottom of the parabola.

1.10 Nuclear Reactions

As chemical reactions alter the distribution of the outer electrons of the atoms, giving rise to the formation of molecules, reactions between two nuclei can give rise to the formation of nuclei different from the ones that began the process.

For example, when a helium and a nitrogen nucleus collide, the 18 nucleons constituting them can rearrange in such a way that the nucleus ^{17}O is formed and a proton is liberated. Similarly to chemical reactions, this process is denoted as

$$^4\text{He}_2 + {}^{14}\text{N}_7 \rightarrow {}^{17}\text{O}_8 + {}^1\text{H}_1. \tag{1.10}$$

or

$$\alpha + {}^{14}\text{N}_7 \rightarrow {}^{17}\text{O}_8 + p. \tag{1.11}$$

Another notation commonly used is the following

$$^{14}\text{N}(\alpha, p)^{17}\text{O}, \tag{1.12}$$

where the target nucleus is written first and then a parenthesis followed by the resulting nucleus. The parenthesis contains the projectile particle first, followed by the comma and then the lighter escaping particle (or particles).

Writing equations like (1.10), one should always check the superscripts and subscripts of the nuclei in order to have the same total number of nucleons and the same electric charge on both sides of the equation. In the above example, we have in total 18 nucleons and positive electric charge $+9e$ in both the initial and final states of the reaction.

As for chemical reactions, also nuclear reactions can be either *exothermic* (i.e. releasing energy) or *endothermic* (i.e. requiring an energy input). In order to know if a nuclear reaction $X(a,b)Y$ releases or absorbs energy, we only need to compare the total masses on the left and right sides of the equation. If the total rest-mass multiplied by c^2 of the left side minus that on the right side of the equation is positive, the reaction is exothermic; if it is negative, the reaction is endothermic.

It is convenient to introduce the Q-value of a reaction, which measures the energy gained (or lost) due to the difference between the initial and final masses. Using energy conservation for the reaction $X(a,b)Y$, in which a particle a hits a nucleus X at rest, gives

$$M_a c^2 + T_a + M_X c^2 = M_b c^2 + T_b + M_Y c^2 + T_Y, \tag{1.13}$$

where T represents the kinetic energy of each particle. The masses of a and X are the ground state masses. On the other hand, the collision may leave the nucleus Y in excited states; in that case, $M_Y c^2$ represents the total energy of that state. The Q value of the reaction is defined as the difference between the final and initial kinetic energies

$$Q = T_b + T_Y - T_a, \tag{1.14}$$

which, from Eq. (1.13) can be also written

$$Q = [(M_a + M_X) - (M_b + M_Y)] c^2. \tag{1.15}$$

When Q is positive, the reaction is exothermic, or *exoergic*, since it releases kinetic energy by the conversion of a portion of the rest mass into kinetic energy. Conversely, when Q is negative, the reaction is endothermic, or *endoergic*: the Q-value is numerically equal to the kinetic energy converted into rest mass in the reaction. In this case, an energy threshold exists, $(T_a)_{th,}$ for the kinetic energy of the incident particle a below which the reaction cannot take place. The threshold energy is[7]

$$(T_a)_{th} = -Q\left(1 + \frac{M_a}{M_X}\right). \qquad (1.16)$$

The energies at play in nuclear reactions are much greater (about a million times) than those involved in chemical reactions. Therefore, special situations are needed to make nuclear reactions to occur. Such conditions occur, for example, in the central region of the stars, in particle accelerators or nuclear reactors.

The reaction in Eq. (1.10) was the first man-made reaction to be observed. It was the New Zealand-born British physicist Ernest Rutherford (Brightwater, New Zeeland, 1871, Cambridge, England, 1937) to demonstrate, in 1919, that alpha particles produced from the decay of bismuth-214 could knock protons out of nitrogen nuclei and merge with what was left behind. In 1925, the British physicist Patrick M.S. Blackett (London, England, 1897, London, England, 1974) succeeded in photographing examples of this reaction in a suitable detector (*cloud chamber*).

Today, nuclear reaction experiments are no longer performed with alpha particles from radioactive sources. Large accelerators produce beams of a wide variety of particles, ranging from electrons and protons to uranium nuclei. The energy of these projectiles is very much higher than alpha particles emitted in radioactive decay and can be precisely controlled. In addition, the number of particles per second is vastly greater than a radioactive source could possibly produce. Beams of high energy particles can be directed onto targets containing any chosen (typically stable) nuclei, where they collide and nuclear reactions take place. When two nuclei collide at high energies, many different processes occur, leading to the production of a variety of final products.

After seven decades of research using particle accelerators and nuclear reactors, almost 3200 isotopes are known in nuclear physics [8]. This number comprises both stable isotopes (some of which known to mankind for centuries), naturally occurring radioactive isotopes (often called radioisotopes) unveiled after the discovery of radioactivity, and finally radioisotopes that do not occur in nature but are produced in nuclear reactors and at accelerators. The radioactivity of the latter newly discovered nuclei is of the same nature as that displayed by the naturally occurring radioisotopes. All of them, stable, naturally occurring unstable, and artificially produced unstable isotopes, are generally listed in an orderly fashion in what is known as an *isotope chart* or *chart of nuclides*.

---

[7]Equation (1.16) is an excellent approximation (namely, $Q \to 0$, or $M_a + M_X \cong M_b + M_Y$) of the correct relation $(T_a)_{th} = -Q \times \frac{(M_a + M_X + M_b + M_Y)}{2M_X}$.

Problem 1.13: Nuclear reactions. Determine the nature of the nucleus X produced in the reaction $^{10}B_5(n,\alpha)^A X_Z$.
Solution: In the reaction, the numbers of protons and neutrons are conserved; therefore one has:

$$\text{Proton number}: \quad 5 = Z + 2;$$
$$\text{Neutron number}: \quad 5 + 1 = N + 2;$$

which gives $Z = 3$ and $N = 4$. Then, the residual nucleus is lithium-7, $^7Li_3$.

Problem 1.14: Q-value of reactions. Calculate the Q-value of the reactions $^7Li(p,\alpha)^4He$ and $^4He(\alpha, p)^7Li$. The atomic mass of ^7Li is 7.016005 u; those of the other nuclei can be found in Table 1.3.
Solution: From Eq. (1.15), the Q-value of the reaction $^7Li(p,\alpha)^4He$ is

$$Q_1 = \left[(M_p + M_{Li}) - (M_\alpha + M_{He})\right] c^2$$
$$= [1.007825 + 7.016005 - 4.002603 - 4.002603] c^2 \times (931.494\,\text{MeV}/c^2)$$
$$= 17.35\,\text{MeV}.$$

Clearly, for the inverse reaction the Eq. (1.14) gives

$$Q_2 = -Q_1 = -17.35\,\text{MeV}.$$

Problem 1.15: Q-value of a (d, p) reactions. The effect of a (d, p) reaction is to add a neutron to the target nucleus $^A X_Z$ forming a nucleus $^{A+1} X_Z$. Show that the binding energy of the last neutron in the produced nucleus is given by the sum of Q value of the reaction and the binding energy of the deuteron.
Solution: From Eq. (1.15), the Q-value of the reaction $^A X_Z(d,p)^{A+1} X_Z$ is

$$Q = \left[M(A,Z) + M_d - M(A+1,Z) - M_p\right] c^2.$$

The binding energy b of the last neutron of the nucleus $^{A+1} X_Z$ is (see Problem 1.8)

$$b = \left[M(A,Z) + M_n - M(A+1,Z)\right] c^2.$$

The binding energy of the deuteron $B(^2H)$ is

$$B(^2H) = \left[M_p + M_n - M_d\right] c^2.$$

Summing the first and the last of these equations, one obtains

$$Q + B(^2H) = \left[M(A,Z) + M_d - M(A+1,Z) - M_p\right] c^2 + \left[M_p + M_n - M_d\right] c^2 = b.$$

1.11   Nuclear Abundance

Chemical elements are formed through nuclear reactions inside stars and in stellar explosions. During this process, known as nucleosynthesis, a multitude of different types of isotopes are formed. Most of these are unstable and decay into stable nuclei either directly or via several intermediate steps.

The elements up to iron are produced by fusion reactions inside stars. Beginning with the fusion of hydrogen into helium, larger and larger nuclei are formed. These processes release energy, which is the reason why the sun shines and provides us with heat. Fusion ceases with the elements in the iron region. This is because fusion into larger nuclei would require energy input. As a result, stars begin to burn out when this stage has been reached.

Nuclei heavier than iron are produced by neutron or proton capture reactions[8] at the end of the lives of large stars—so-called *red giants* and in violent explosions of stars. In the phase of the explosion, all nuclei heavier than iron and many nuclei of intermediate mass are produced.

One such event, that occurred approximately 168,000 years ago in the Magellan Cloud (which is situated 168,000 light-years from Earth), which is visible in the southern hemisphere of the Earth, was observed in February 23, 1987.

Table 1.4 gives the fifteen most abundant elements in the Universe. The data are estimates of the average composition of the universe [13]. They have been rounded for convenience; this is why their sum is slightly higher than 100. Hydrogen is by far the most abundant element followed by helium, which is also quite abundant, and, with much smaller percentages, by heavier elements such as oxygen, carbon, neon, iron, and nitrogen. Together, hydrogen and helium account for about 98 % of all elements in the Universe.

The Fig. 1.12 shows the abundance of the chemical elements in the Solar system, normalised to the abundance of silicon, plotted as a function of their atomic number Z on a logarithmic scale that spans twelve orders of magnitude, or thousands of billions. Values given here are from Ref. [13]. Data given in different sources vary somewhat (mainly for the rarest isotopes), reflecting the difficulty in assessing these numbers.

As indicated above, hydrogen and helium are the most common elements. Then the distribution falls with increasing Z, showing peaks for elements whose most abundant isotope corresponds to the mass number $A = 4n$, with n integer: ^{4}He, ^{12}C, ^{16}O, ^{20}Ne, ^{24}Mg, ^{28}Si, ^{32}S, ^{36}Ar and ^{40}Ca, all comprising an integer number of ^{4}He nuclei. A prominent peak is also seen at ^{56}Fe, which is believed to be the result of the decay of radioactive ^{56}Ni produced in the last stage of the stellar nuclear burn up cycle, starting from fourteen ^{4}He nuclei.

---

[8]In capture reactions, either a neutron or a proton stick to a nucleus forming a new nuclide. Since neutrons are electrically neutral, their capture can occur simply by effect of the nuclear force even when the neutron is very slow. Protons have to overcome the electric repulsion and so, to be captured, they have to possess sufficient kinetic energy.

Table 1.4 Relative abundance of top fifteen elements in the Universe [13]

Percentage of the most abundant elements in the Universe (%)	
Hydrogen	75.0
Helium	23.0
Oxygen	1.0
Carbon	0.5
Neon	0.12
Iron	0.10
Nitrogen	0.10
Silicon	0.06
Magnesium	0.05
Sulphur	0.05
Argon	0.02
Calcium	0.007
Aluminium	0.005
Nickel	0.004
Sodium	0.002

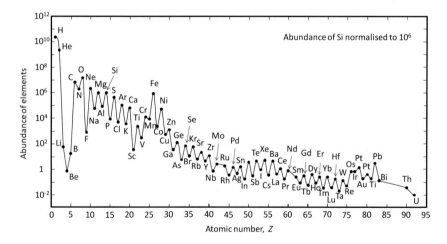

Fig. 1.12 Estimated abundance of the chemical elements in the solar system as a function of their atomic number Z, normalised to the abundance of silicon (= 10^6). The elements just above iron are somewhat common (being the easiest to make), and then the abundances further up are much lower. Obtained with data from Ref. [13]

Beyond the iron peak, the distribution continues to fall with increasing Z, showing peaks for elements whose most abundant isotope corresponds to the mass number A = 80, 87, 130, 138, 195, 208. These peaks are due to the higher probability of neutron captures responsible for the production of many heavy elements, leading to these specific isotopes.

Problem 1.16: The life of the Sun. The production of solar energy is a case
of continuous transformation of mass into energy. The reactions that give rise
to this process are fusion thermonuclear reactions of light elements. The production of the deuterium nucleus is the first process of a cycle (see Problem 1.4)
that leads, through intermediate stages, to the formation of a nucleus of ^{4}He by
the combination of four protons, releasing two positrons, two neutrinos and two
photons and an energy equal to 24.7 MeV (one can easily calculate it knowing
that the mass of helium is $M_{He} = 6.64466 \times 10^{-27}$ kg, and using the values of
the proton and electron masses given in Table 1.1). Knowing that every second
on each square metre of the Earth arrives, in the form of solar radiation, an
amount of energy equal to 1360 J (i.e. the solar power flux on Earth is
$P_S = 1.36$ kW/m^2), calculate the "fuel" actually consumed per second on the
Sun.
Solution: The Sun emits energy in all directions; then the same amount of
energy falls on each square metre of a hypothetical spherical surface centred
on the Sun and with a radius R equal to the Earth-Sun distance
($R = 1.5 \times 10^{11}$ m). Therefore, the total power radiated by the Sun is:

$$P = 4\pi R^2 P_S = 4 \times 3.14 \times \left(1.5 \times 10^{11}\right)^2 \times 1.36 = 3.84 \times 10^{23} \text{ kW}$$
$$= \left(\text{remembering that } 1 \text{ J} = 6.25 \times 10^{12} \text{MeV}\right) = 2.40 \times 10^{39} \text{ MeV/s}.$$

Clearly, the energy radiated per second is $E = 2.40 \times 10^{39}$ MeV.
Applying the calculation outlined in the Problem 1.6 to the fusion of four
initial protons into a ^{4}He nucleus plus two positrons, one finds that

$$\frac{4M_p - M_{He} - 2m_e}{M_{He}} = \frac{4 \times 1.67262 \times 10^{-27} - 6.64466 \times 10^{-27} - 2 \times 9.10938 \times 10^{-31}}{6.64466 \times 10^{-27}}$$
$$= \frac{4.3998 \times 10^{-29}}{6.64466 \times 10^{-27}} = 0.0066,$$

that is, the energy E corresponds to 0.66 % of the mass M involved in the
fusion reactions in a second, $E = 0.0066 \, Mc^2$, from which it follows:

$$M = \frac{2.40 \times 10^{39}}{0.0066} \cong 3.63 \times 10^{41} \text{ MeV/c}^2.$$

Taking into account that the mass of the proton is about 938 MeV/c^2, every
second the Sun burns $N_p = M/938 \cong 3.87 \times 10^{38}$ protons, which corresponds
to more than six hundred million tonnes of hydrogen per second
($3.87 \times 10^{38} \cdot 1.67 \times 10^{-27} \cong 6.46 \times 10^{11}$ kg/s). This amount of mass,
although huge, is only a tiny fraction ($\approx 3 \times 10^{-19}$) of the mass of the Sun
($M_S = 1.99 \times 10^{30}$ kg).

Problems

1-1 Tellurium-125 (^{125}Te) is a stable isotope of tellurium used for biological and biomedical labelling, as target materials and other applications. Calculate the radius of the isotope ^{125}Te, its volume and density, knowing that its atomic mass is 124.904431 u).
[*Ans.*: 6.0×10^{-15} m; 9.04×10^{-43} m^3; 2.29×10^{17} kg/m^3]

1-2 Neutron stars are composed almost entirely of neutrons condensed at the density of nuclear matter, 1.8×10^{17} kg/m^3. They are the densest and smallest stars known to exist in the Universe, with a radius of only about 12–14 km. Calculate the radius of a neutron star that has the mass of the Sun (1.989×10^{30} kg).
[*Ans.*: 13.8 km]

1-3 The nuclide ^{62}Ni$_{28}$ is the most tightly bound of any known nuclide. Calculate its binding energy per nucleon, knowing that its atomic mass is 61.928345 u.
[*Ans.*: 8.794 MeV]

1-4 The isotope ^{88}Sr$_{38}$ is the most abundant of the four naturally occurring isotopes of strontium. It is a stable isotope with 50 neutrons. Calculate the binding energy per nucleon of the ^{88}Sr$_{38}$, knowing that its atomic mass is 87.905612 u).
[*Ans.*: 8.73 MeV]

1-5 What is the nucleus $^Y Z_X$ that is produced in the reaction $p + {}^{59}Co_{27} \rightarrow n + {}^Y Z_X$?
[*Ans.*: Nickel-59, ^{59}Ni$_{28}$]

1-6 Evaluate the nature of the nucleus X produced in the following reactions:
(a) $n + {}^{16}O_8 \rightarrow X$.
(b) $\alpha + {}^{118}Sn_{50} \rightarrow X + n$.
(c) $p + {}^{127}I_{53} \rightarrow {}^{50}Sc_{21} + X$.
(d) $n + {}^{235}U_{92} \rightarrow {}^{107}Tc_{43} + X + 5n$.
[*Ans*: (a) oxygen-17, ^{17}O$_8$; (b) tellurium-121, ^{121}Te$_{52}$; (c) arsenic-78, ^{78}As$_{33}$; (d) indium-124, ^{124}In$_{49}$]

1-7 Calculate the Q-value of the reaction ^{4}He(^{3}He,γ)^{7}Be, knowing that the atomic mass of helium-3 is 3.016029 u and that of beryllium-7 is 7.016930 u.
[*Ans.*: 1.58 MeV]

1-8 Calculate the Q-value of the so-called "triple-alpha process", ^{4}He + ^{4}He + ^{4}He → ^{12}C, by which three helium-4 nuclei are transformed into carbon. This process occurs in the later phase of red giants and red supergiants, when the central temperature of the stars is high enough.
[*Ans.*: 7.274 MeV]

1-9 Calculate the energy released in the oxygen-burning reaction ^{16}O$_8$ + ^{16}O$_8$ → ^{28}Si$_{14}$ + ^{4}He$_2$, which occurs in the hot core of large stars, knowing that the atomic mass of the isotope silicon-28 is 27.976926 u.
[*Ans.*: 9.59 MeV]

1-10   Determine the Q-value of the reaction $^{14}N(d,p)^{15}N$, knowing that the mass of
       the nucleus nitrogen-15 is 15.000109 u.
       [*Ans.*: Q = 8.609 MeV]
1-11   The reactions $^{207}Pb(d,p)^{208}Pb$ and $^{208}Pb(d,p)^{209}Pb$ have Q-values of 5.14
       and 1.64 MeV, respectively. What are the binding energies of the last neu-
       tron in ^{208}Pb and ^{209}Pb?
       [*Ans*: 7.36 MeV; 3.86 MeV]
1-12   Calculate the threshold energy of the reaction: $\gamma + {}^{12}C_6 \rightarrow {}^4He_2 + {}^4He_2 + {}^4He_2$.
       [*Ans.*: 7.27 MeV]
1-13   Show that the ^8Be nucleus (atomic mass 8.005305 u) is unstable for breaking
       into two alpha particles.
       [*Ans.*: Q = 0.09 MeV]

References

1. INFN-Nuclei e stelle, Asimmetrie, **9**, 7 (Sept 2009). http://www.asimmetrie.it/index.php/al-cuore-della-materia
2. CODATA (Committee on Data for Science and Technology), Internationally recommended 2014 values of the fundamental physical constants. http://physics.nist.gov/cuu/Constants/index.html
3. D. Mendeleev, Zeitschrift für Chemie **12**, 405 (1869)
4. Pure Appl Chem. **85**, 1047 (2013). http://www.chem.qmul.ac.uk/iupac/AtWt
5. Jefferson Lab, http://educational.jlab.org/itselemental/tableofelements.pdf
6. R. Hofstadter, Ann. Rev. Nucl. Sci. **7**, 231 (1957)
7. I. Sick, Nucl. Phys. A **218**, 509 (1974)
8. http://periodictable.com/Properties/A/KnownIsotopes.html
9. G. Fea, Il Nuovo Cimento **2**, 368 (1935)
10. Nucleus—A Trip into the Heart of Matter, a PANS (Public Awareness of Nuclear Science) book. (2002), Chap. 6, p. 71. http://www.nupecc.org/pans/bookcontent.html
11. G. Audi, A.H. Wapstra, Nucl Phys **A565**, 1 (1993) and Nucl Phys **A595**, 409 (1995)
12. The University of Liverpool, Department of Physics, http://ns.ph.liv.ac.uk/∼ajb/radiometrics/neutrons/images/beta_stability_valley.gif
13. http://periodictable.com/Properties/A/UniverseAbundance.an.log.html

Chapter 2
Radioactivity and Penetrating Power of Nuclear Radiation

In this chapter, we primarily discuss radioactivity, the process by which an unstable nucleus loses energy by emitting radiation. First we illustrate the main types of decays (alpha, beta, gamma, internal conversion, electron capture, nucleon emission and fission), then we discuss the radioactive decay law, decay rates, mean lifetimes of isotopes, radioactive families and sequential decays.

In the second part of the chapter, we discuss the interaction between radiation and matter, the existence of an artificial radioactivity in addition to natural one, due to the development of nuclear technologies, the biological effects of ionising radiation with elements of dosimetry, and the applications of nuclear radiation in medicine, research and industry.

2.1 Nuclear Decay

As discussed in the previous Chapter, a nucleus containing a number of protons and neutrons not well balanced is unstable. It will break spontaneously—that is, without supplying external energy—in pieces or rearrange its internal structure with a release of energy. This phenomenon is called *radioactivity*. The unstable nucleus is said to be *radioactive* and the change is called *radioactive decay*. The original nucleus which decays is called the *parent nuclide*, and the product nucleus is called the *daughter nuclide*. When the parent and the daughter nuclides are different chemical elements, the decay process results in the creation of an atom of a different element. This process is known as *a nuclear transmutation*. In the decay, both the total electric charge and the number of nucleons are conserved, i.e. must remain the same before and after the decay.

There are many types of radioactive decay (see Table 2.1). A radioactive decay is classified as α-, β-, or γ-decay, according to the type of particle emitted. Other types of decay produce emissions of nucleons or nuclei lighter than the parent nucleus. All these radiation emissions are determined and measured with great precision and are typical of each radioactive nucleus.

© Springer International Publishing Switzerland 2016
E. De Sanctis et al., *Energy from Nuclear Fission*, Undergraduate Lecture Notes in Physics, DOI 10.1007/978-3-319-30651-3_2

Table 2.1 Main features of the different types of radioactivity

Radioactivity	Change in nuclear charge	Change in mass number	Characteristic of the process
α-decay	$Z - 2$	$A - 4$	Emission of an alpha particle ($A = 4$, $Z = 2$) from a nucleus
β⁻-decay	$Z + 1$	A	A neutron in a nucleus transforms into a proton plus an electron and an antineutrino: $n \rightarrow p + e^- + \bar{\nu}$
β⁺-decay	$Z - 1$	A	A proton into a nucleus transforms into a neutron plus a positron and a neutrino: $p \rightarrow n + e^+ + \nu$
Electron capture	$Z - 1$	A	A nucleus captures an atomic electron and emits a neutrino. The daughter atom is left in an excited and unstable state. $p + e^- \rightarrow n + \nu$
γ-decay	Z	A	A nucleus in an excited state makes a transition to a state with lower energy by emitting a photon
Internal conversion	Z	A	An excited nucleus transfers energy to an atomic electron, which is subsequently ejected from the atom
n emission	Z	$A - 1$	Emission of a neutron from the nucleus
p emission	$Z - 1$	$A - 1$	Emission of a proton from the nucleus
Spontaneous fission	$\sim \frac{1}{2} Z$	$\sim \frac{1}{2} A$	The nucleus disintegrates, usually into two fragments of very approximately equal mass and charge (see Chap. 3)

2.1.1 α-Decay

For the heaviest nuclei whose average nuclear binding energy is relatively low, some of their nucleons can go towards a higher binding energy by emitting a helium nucleus, $^4\text{He}_2$, which is called α-particle in radioactive decays. An α-particle has a double positive charge $+2e$, and mass number $A = 4$. Then, the daughter nucleus has a mass number A and an atomic number Z lower than the parent by four and two units, respectively.

As we have seen in Sect. 1.8, the helium nucleus is a cluster of four nucleons—two protons and two neutrons—extraordinarily strongly bound (binding energy per nucleon $B/A = 7.06$ MeV/nucleon, a very high value for such a light nucleus). Then it is not surprising that all heavy nuclei ($A \geq 180$) are energetically unstable against α-decay. Specifically, α-decay is energetically allowed if the mass of the parent nucleus is larger than the sum of those of the daughter and helium nuclei:

$$M(Z, A) - M(Z - 2, A - 4) - M(^4\text{He}) > 0. \qquad (2.1)$$

This relation can be written in terms of the total binding energies of the nuclei, obtaining

$$B(^4\text{He}) > B(Z, A) - B(Z - 2, A - 4). \qquad (2.2)$$

Therefore, in α-decay the parent nucleus releases some energy Q_α. This is given by the mass difference between the parent and the daughter nucleus multiplied by c^2:

$$Q_\alpha = \left[M(Z,A) - M(Z-2, A-4) - M\left(^4\text{He}\right)\right]c^2. \qquad (2.3)$$

The energy Q_α goes into the kinetic energies of the α-particle and the daughter nucleus (see Problem 2.1).

An example of α-decay is the transmutation of uranium-238 into thorium-234:

$$^{238}\text{U}_{92} \rightarrow{}^{234}\text{Th}_{90} + {}^4\text{He}_2.$$

The thorium nucleus has 2-protons and 2-neutrons less that the uranium: the missing nucleons are taken away by the α-particle.

Problem 2.1: Alpha decay of ^{238}U. Calculate the energy released in the α-decay of uranium-238 (rest mass M_U = 238.0508 u) into thorium-234 (rest mass M_{Th} = 234.0436 u). Calculate the fraction of this energy taken by the α-particle.

Solution: With clear meaning of the symbols we have

$$\Delta m = M_U - M_{Th} - M_{He} = 238.0508\,\text{u} - 234.0436\,\text{u} - 4.0026\,\text{u} = 0.0046\,\text{u}.$$

$$Q = \Delta m\,c^2 = 0.0046 \times 931.494 = 4.28\,\text{MeV}.$$

This energy is shared between the thorium nucleus and the α-particle. The nucleus ^{234}Th has a mass much larger than that of the α-particle; therefore it recoils with smaller speed. In the decay both the energy and the linear momentum are conserved, then assuming the parent nucleus to be at rest and neglecting relativistic corrections, one has:

$$M_{Th}v_{Th} = M_\alpha v_\alpha,$$

$$\frac{1}{2}M_{Th}v_{Th}^2 + \frac{1}{2}M_\alpha v_\alpha^2 = k_{Th} + k_\alpha;$$

where we have indicated with k_α and k_{Th} the kinetic energies of the helium and thorium nuclei, respectively. Clearly, $k_{Th} + k_\alpha = Q$.

From the first relationship we get $v_\alpha = \frac{M_{Th}}{M_\alpha}v_{Th}$, and then, from the second relationship, the kinetic energy of the α-particle is

$$k_\alpha = \frac{1}{2}M_\alpha v_\alpha^2 = \frac{1}{2}M_\alpha\left(\frac{M_{Th}}{M_\alpha}v_{Th}\right)^2 = \frac{1}{2}M_{Th}v_{Th}^2\frac{M_{Th}}{M_\alpha} = \frac{M_{Th}}{M_\alpha}k_{Th} = \frac{234.0436}{4.0026}k_{Th}$$

$$= 58.47\,k_{Th}.$$

Then the energy Q released in the decay is equal to:

$$Q = k_{Th} + k_\alpha = 58.47\,k_{Th} + k_{Th} = 59.47\,k_{Th}.$$

From which it results

$$k_{Th} = \frac{Q}{59.47} = 0.017\,Q = 0.073\,\text{MeV} = 73\,\text{keV},$$

$$k_\alpha = 58.47\,k_{Th} = \frac{58.47}{59.47}\,Q \approx 0.983\,Q = 4.21\,\text{MeV}.$$

The α-particle, which is much lighter than thorium, takes $\approx 98.3\,\%$ of the total released energy. The thorium nucleus takes the remaining 1.7 %.

2.1.2 β^--Decay

A β^--decay (*beta-minus decay*) occurs when the ratio of neutrons to protons in the nucleus is too high. In this case, an excess neutron transforms into a proton and an electron plus a neutral particle of negligible mass, called *antineutrino* (symbol $\bar{\nu}$). The proton stays in the nucleus and the electron is ejected. In the process, the parent nucleus transforms into a daughter one having the same mass number A, and the atomic number Z greater by one unit:

$$^A X_Z \rightarrow\,^A X_{Z+1} + e^- + \bar{\nu}.$$

Examples of β^--decay are those of nuclei up the left side of the parabola of Fig. 1.10, namely of niobium-101 into molybdenum-101, molybdenum-101 into technetium-101, and technetium-101 in ruthenium-101:

$$^{101}Ni_{41} \rightarrow\,^{101} Mo_{42} + e^- + \bar{\nu},$$
$$^{101}Mo_{42} \rightarrow\,^{101} Tc_{43} + e^- + \bar{\nu},$$
$$^{101}Tc_{43} \rightarrow\,^{101} Ru_{44} + e^- + \bar{\nu}.$$

Energetically, β^--decay is possible whenever the mass of the daughter atom $M(A, Z + 1)$ is smaller than the mass of its isobaric neighbour:

$$M(A,Z) > M(A,Z+1). \tag{2.4}$$

We consider here the atomic masses and so the rest mass of the electron created in the decay is automatically taken into account. The tiny mass of the antineutrino

(<2 eV/c^2 [1]) is negligible in the mass balance. The decay energy of a β^--decay is given by the mass difference between the parent and the daughter atom multiplied by c^2:

$$Q_{\beta^-} = [M(A,Z) - M(A,Z+1)]c^2. \qquad (2.5)$$

Problem 2.2: Beta decay of ^{234}Th. Calculate the maximum energy of the electron emitted in the β^--decay of thorium-234, ^{234}Th$_{90}$ (rest mass $M_{Th} = 234.04359$ u), into protoactinium-234, ^{234}Pa$_{91}$ (rest mass $M_{Pa} = 234.04330$ u).

Solution: The atom ^{234}Pa$_{91}$ produced by the β^--decay of ^{234}Th$_{90}$ is not neutral as its nucleus has one proton more ($Z = 91$) than the parent nucleus, but the same number of bound electrons (90). Then it has one electron less. However, if we sum up the mass of the produced Pa and that of the emitted β^- particle, we get the mass of the Pa nucleus plus the mass of 91 electrons, which is exactly the atomic mass of the neutral Pa atom (except for the negligible binding energy of the atomic electron). Then the mass difference Δm corresponding to the β^--decay is

$$\Delta m = M_{Th} - M_{Pa} = 234.04359\,u - 234.04330\,u = 0.00029\,u.$$

The equivalent energy released in the decay is

$$Q = \Delta mc^2 = (0.00029\,u) \times (931.494\,MeV/c^2) = 0.27\,MeV.$$

This energy is redistributed among the daughter nucleus, the electron and the antineutrino. From the exact equations of the reaction kinematics, it is possible to show that, to a very good approximation, the maximum electron energy is given by the total energy released in the reaction. Hence, 0.27 MeV is, to a good approximation, the maximum energy that the emitted electrons can have.

N.B. The neutrino emitted in the decay does not contribute to the mass balance, as its mass is negligible.

2.1.3 β^+-Decay

A β^+-decay (*beta-plus decay*) occurs when a proton inside a nucleus converts into a neutron by emitting a positron (i.e., a positive electron) and a neutral particle called *neutrino* (symbol v). In such a process, the parent nucleus transforms into a daughter nucleus having the same mass number A and the atomic number Z smaller by one unit:

$$^A X_Z \rightarrow {}^A X_{Z-1} + e^+ + \nu.$$

Examples of β^--decay are those of nuclei up the right side of the parabola of Fig. 1.10, namely of silver-101 in palladium-101, palladium-101 in rhodium-101, and rhodium-101 in ruthenium-101:

$$^{101}Ag_{47} \rightarrow {}^{101}Pd_{46} + e^+ + \nu,$$
$$^{101}Pd_{46} \rightarrow {}^{101}Rh_{45} + e^+ + \nu,$$
$$^{101}Rh_{45} \rightarrow {}^{101}Ru_{44} + e^+ + \nu.$$

Energetically, β^+-decay is possible whenever the following relationship between the masses of the parent and daughter atoms, $M(A, Z)$ and $M(A, Z-1)$, is satisfied:

$$M(A, Z) > M(A, Z-1) + 2m_e. \tag{2.6}$$

This relationship takes into account that the final system of a β^+ decay consists of a nucleus $(Z-1, A)$ with $(Z-1)$ atomic electrons, together with the emitted positron as well as of the existence of an excess electron in the parent nucleus. Hence, with clear meaning of the symbols, the total mass of the final system is $M(A, Z-1) + m_{\beta+} + m_{e-}$, that is, recalling that the positron has the same mass as the electron $(m_{\beta+} = m_{e-} = m_e)$, $M(A, Z-1) + 2m_e$.

Then, the total energy available for the β^+-decay is

$$\begin{aligned} Q_{\beta+} &= \{M(A, Z) - [M(A, Z-1) + 2m_e]\}c^2 \\ &= [M(A, Z) - M(A, Z-1) - 2m_e]c^2 = \Delta m c^2 - 2m_e c^2. \end{aligned} \tag{2.7}$$

Therefore, the maximum kinetic energy of the emitted positron is $2m_e c^2$ less than the mass-energy difference $\Delta m c^2$ between the parent and daughter nucleus.

2.1.4 Electron Capture

Electron capture is a decay process in which an atomic electron interacts with the nucleus, where it combines with a proton, forming a neutron and a neutrino. The neutrino is ejected from the atomic nucleus:

$$p + e^- \rightarrow n + \nu.$$

During electron capture, an atom changes a proton into a neutron, so that it changes from one element to another. However, the total number of nucleons (protons + neutrons) remains the same. For example, after undergoing electron capture, an atom of beryllium (with 4 protons and 3 neutrons) becomes an atom of lithium (with 3 protons and 4 neutrons):

$$^7\text{Be}_4 + e^- \rightarrow {}^7\text{Li}_3 + \nu.$$

Electron capture results in a neutral atom, since the loss of one atomic electron is balanced by the loss of one positive nuclear charge. The process leaves a vacancy in the electron energy level from which the electron came, and that vacancy is filled by the dropping down of a higher-level electron with the release of some energy. Although most of the time this energy is released in form of an electromagnetic radiation, called X-ray, having a wavelength of the order of 10^{-8} to 10^{-11} m, such energy can also be transferred to another atomic electron, which is ejected from the atom. This second electron is called an *Auger electron* and the process called *Auger effect* after one of its discoverers.[1] In the latter case, a positive ion is left over.

Electron capture reactions compete with β^+-decay. It can occur when the following energy conservation condition is satisfied

$$M(A,Z)c^2 > M(A, Z-1)c^2 + \epsilon,$$

where ϵ is the excitation energy of the atomic shell of the daughter nucleus. The condition stems from the fact that the atom of the new element formed in the reaction has a vacancy in an electron energy level, i.e. it is produced in an excited atomic energy level. Clearly, the energy available in the capture process, given by the difference $[M(A, Z) - M(A, Z - 1)]c^2$, must be greater than ϵ.

2.1.5   γ-Decay

The nucleus does not have a rigid structure and may undergo excitations and oscillations accompanied by changes of shape. A γ-decay (gamma decay) occurs when a nucleus that is in an excited state (that is, in a configuration with energy higher than the ground state) makes a transition to either an excited state with lower energy than the initial one or to the ground state (we recall that this is the most stable configuration of the nucleus). The energy resulting from the jump from one level up to the other level down is emitted as an electromagnetic wave, called γ-ray, having a wavelength of the order of 10^{-10} to 10^{-13} m, without other particles being emitted in the decay. Gamma rays are also described as packets of energy, called *photons*.

Both the number of protons and neutrons in the nucleus do not change in this process as photons are electrically neutral, so the parent and daughter atoms are the same chemical element.

---

[1]Pierre Victor Auger (Paris, France, 1899—Paris, France, 1993) was a French physicist who has made important contributions in the fields of atomic, nuclear, and cosmic ray physics.

An example of a γ-decay is the decay of excited radon-222

$$^{222}\text{Rn}_{86}^* \rightarrow {}^{222}\text{Rn}_{86} + \gamma.$$

The unstable parent nucleus $^{222}\text{Rn}_{86}^*$ (the asterisk is used to indicate that the nucleus is not in its energy ground state) is itself the daughter nucleus of a preceding α-decay.

2.1.6 Internal Conversion

Internal conversion is a radioactive decay process in which an excited nucleus, instead of emitting a gamma ray, de-excites by knocking out one of the electrons in the atom. The electron is emitted with a well-defined energy, the same energy that a gamma ray would have had in the same decay process (as usually, we neglect the atomic binding energy). Thus, in an internal conversion process, a high-energy electron is emitted from the radioactive atom, not from the nucleus. For this reason, the high-speed electrons resulting from internal conversion are not beta particles, since the latter come from beta decay, where they are produced in the nuclear decay process.

During internal conversion, the atomic number does not change, and thus (as is the case of gamma decay) no transmutation of one element to another takes place. However, since an electron is lost, a hole appears in an electron shell which is subsequently filled by an upper electron, producing an X-ray or an Auger electron.

2.1.7 Nucleon Emission

Highly excited neutron-rich or proton-rich nuclei, formed as the product of other types of decay, can occasionally lose energy by way of neutron and proton emission, resulting in a transition from one isotope to another of the same element, or from a nuclide of one element to a nuclide of another element, respectively.

2.1.8 Spontaneous Fission

In another type of radioactive decay, the nucleus disintegrates into other nuclei that are not well defined, but rather correspond to a range of fragments of the original nucleus. This decay, called spontaneous fission (see Chap. 3), happens when a large unstable nucleus spontaneously splits into two (and occasionally three) smaller daughter nuclei, and generally leads to the emission of gamma rays, neutrons, and beta particles.

2.1.9 Summary

Table 2.1 summarises the main features of the above different types of radioactivity.

Each unstable nuclide emits radiation of the same type and with the same total energy; for example, a γ emitter always emits γ-rays of certain fixed energies.

If a radioactive nucleus decays to another unstable nucleus, then that will also decay. Radioactive decay will continue until a stable nucleus is formed. This is the case, for example, of uranium-238, which transmutes into lead with a sequence of fourteen decays (see Sect. 2.3).

Alpha decay mainly occurs in the nuclides of the heavier elements (the lightest nuclide featuring alpha-decay being ^{105}Te, $Z = 52$ [2]), while beta decay occurs in the nuclides of all elements: precisely the nuclides that are above the stability curve of Fig. 1.7 decay β^+, those found under the curve decay β^-. Gamma decay often accompanies an alpha or beta decay. In fact, when a radioactive nucleus decays, the daughter nucleus is not necessarily produced in the ground state. When the daughter nucleus is produced in an excited state, it decays rapidly to the ground state, either directly, or by subsequent emissions of gamma rays when passing through other intermediate excited states. Gamma rays are, therefore, the most common form in which the surplus energy of the excited products of radioactive decay is liberated.

2.2 The Radioactive Decay Law

The decay events take place completely at random; there is no way of predicting the precise instant at which a particular nucleus will decay: it can happen next moment, next day or even next century. However, we can compute decay probabilities.

Each radioactive isotope has a characteristic, well-defined probability of decaying per unit time, which is denoted by λ and takes the name of *decay constant*. This it is an intrinsic property of the nucleus itself, which is independent of all physical and chemical conditions, such as temperature, pressure, concentration, age or past history of the radioactive nuclide. Radioactive nuclei decay spontaneously because of their internal dynamics and therefore their decay constants never change.

In a specimen containing N radioactive atoms at an arbitrary instant t, the number of decays dN that occur on average in a short time interval dt is proportional to dt and N, namely, according to the definition of λ:

$$dN = -\lambda N dt, \tag{2.8}$$

where the minus sign in the second member indicates that the number N of radioactive nuclei decreases with time. Obviously, being nuclear decay a random phenomenon, the number dN is subject to statistical fluctuations: it represents the average number of decays in the short time interval dt.

Equation (2.8) can be rewritten, separating the variables,

$$\frac{dN}{N} = -\lambda dt,$$

which, by integrating both sides

$$\int_{N_0}^{N} \frac{dN}{N} = -\int_{0}^{t} \lambda dt,$$

where N_0 is the number of radioactive nuclei present at time zero and N represents the number of the nuclides surviving until the time t, gives

$$\ln\frac{N}{N_0} = -\lambda t.$$

Taking the anti-logs of both sides of this equation, we get:

$$N(t) = N_0 e^{-\lambda t}, \tag{2.9}$$

which is called the *radioactive decay law*.

To measure the rate of decay the *half-life* or *half-period*—usually denoted by $\tau_{1/2}$—is often used, which is the time interval over which, on average, half of the radionuclides present in the radioactive material disintegrate.

Equation (2.9) indicates that the number of radioactive nuclei present in a sample decreases with time with the exponential trend shown in Fig. 2.1: after each $\tau_{1/2}$-period there remains half the amount of radionuclides that existed at the beginning of that period. Therefore, after a time of n half-lives the amount of radioactive isotope is reduced to $1/2^n$ of the initial quantity. Again, we have to take the decay law and the halving of the number of radionuclides as representing average behaviours because of the statistical nature of the decay phenomenon.

According to its definition, from the survival Eq. (2.9) it follows that the half-life is given by:

$$\tau_{1/2} = \frac{\ln(2)}{\lambda} = \frac{0.693}{\lambda}. \tag{2.10}$$

Often used is also the *mean-life* of an isotope $\tau = 1/\lambda$, and the previous relation becomes:

$$\tau_{1/2} = 0.693 \times \tau. \tag{2.11}$$

It should be noted that the two variables, mean-life and half-life, have different numerical values. Then, confusing them may result in serious errors.

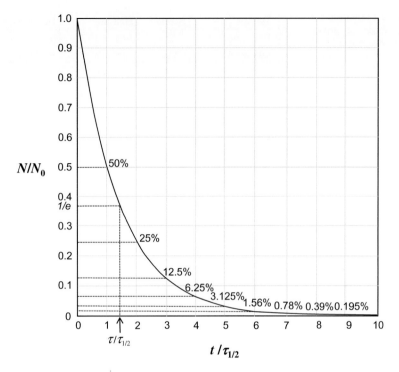

Fig. 2.1 Curve of radioactive decay. The ordinate shows the fraction number of radionuclides surviving until the time $t/\tau_{1/2}$, over an interval of ten half-life periods. After seven half-lives, the amount of radionuclides has been reduced to less than 0.8 % of the initial value. After ten half-lives, to less than 0.1 %

Substitution of τ in Eq. (2.9) shows that the mean-life is the time required for the number of atoms to fall to $e^{-1} = 0.368$ of any initial value.

Table 2.2 shows the half-life and decay mode of some radionuclides. The half-lives of different radioactive nuclei vary within very wide limits, from over one hundred thousand billion years for indium-115, to eight days for iodine-131, down to less than a millisecond for polonium-214. Uranium-238 has a half-life approximately equal to the age of the Earth, which is estimated to be 4.5 billion years. Therefore, about half of the uranium-238 that existed at the time of the formation of the Earth is still present today.

2.2.1   Activity

The amount of radiation emitted by a substance containing radioactive isotopes depends on two factors: the number of unstable nuclei present and their half-life. In order to quantify the amount of radiation emitted, the *activity* dN/dt of a radioactive nuclide is used. This is defined as the number of disintegrations per unit time.

Table 2.2 Half-life and decay modes of certain radioactive nuclei [3]

Isotope	Symbol	Half-life	Decay mode
Polonium-214	$^{214}Po_{84}$	1.64×10^{-4} s	α, γ
Oxygen-15	$^{15}O_8$	2.04 min	β^+, EC
Krypton-89	$^{89}Kr_{36}$	3.15 min	β^-, γ
Nitrogen-13	$^{13}N_7$	9.97 min	β^+, EC
Carbon-11	$^{11}C_6$	20.39 min	β^+, EC
Uranium-239	$^{239}U_{92}$	23.45 min	β^-, γ
Fluorine-18	$^{18}F_9$	109.8 min	β^+, EC
Neptunium-239	$^{239}Np_{93}$	2.36 d	β^-, γ
Radon-222	$^{222}Rn_{86}$	3.82 d	α, γ
Iodine-131	$^{131}I_{53}$	8.03 d	β^-, γ
Tritium	$^3H_1$	12.32 a	β^-
Strontium-90	$^{90}Sr_{38}$	28.79 a	β^-
Cesium-137	$^{137}Cs_{55}$	30.1 a	β^-, γ
Radium-226	$^{226}Ra_{88}$	1600 a	α, γ
Carbon-14	$^{14}C_6$	5700 a	β^-
Americium-243	$^{243}Am_{95}$	7370 a	α, γ
Plutonium-239	$^{239}Pu_{94}$	24,110 a	α, γ
Neptunium-237	$^{237}Np_{93}$	2.14×10^6 a	α, γ
Curium-247	$^{247}Cm_{96}$	1.56×10^7 a	α, γ
Uranium-235	$^{235}U_{92}$	7.04×10^8 a	α, γ
Potassium-40	$^{40}K_{19}$	1.25×10^9 a	β^-, β^+, EC
Uranium-238	$^{238}U_{92}$	4.47×10^9 a	α, γ
Thorium-232	$^{232}Th_{90}$	1.41×10^{10} a	α, γ
Rubidium-87	$^{87}Rb_{37}$	4.81×10^{10} a	β^-
Indium-115	$^{115}In_{49}$	4.41×10^{14} a	β^-

Using Eqs. (2.8) and (2.10) we have:

$$\left|\frac{dN}{dt}\right| = \lambda N = \frac{0.693}{\tau_{1/2}}N, \qquad (2.12)$$

where the absolute value is used because the variation dN is negative.

The unit of activity is the becquerel (Bq), so named in honour of the French physicist Antoine Henri Becquerel (Paris, France, 1852—Le Croisic, France, 1908. Nobel laureate in Physics in 1903), who first discovered the phenomenon of radioactivity. One becquerel is equal to one disintegration or nuclear transformation per second.

A historical unit still widely used is the curie (Ci), so named in honour of the Poland-born French physicist Marie Sklodowska Curie (Warsaw, Poland, 1867—Passy, France, 1934. Nobel laureate in Physics in 1903). One curie is equal to 3.70×10^{10} disintegrations per second. This is the activity of 1.0 g of radium-226. Clearly, 1 Ci = 3.70×10^{10} Bq.

Problem 2.3: Survival of uranium isotopes. The ^{238}U and ^{235}U isotopes have half-lives of 4.47 and 0.7 billion years, respectively. Calculate the fractions of the two isotopes initially present at the formation of the Earth that still exist, knowing that our solar system is considered to be 4.5 billion year old. Calculate the ^{235}U/^{238}U ratio in natural uranium at that date, considering that today it is 0.7 %.

Solution: The ^{238}U isotope has half-life equal to the age of the Earth and then about half of all the ^{238}U initially present still exists today.

For the isotope ^{235}U, the age of the Earth corresponds to more than 6 half-lives. Therefore, there is still only about

$$\frac{1}{2^6} = \frac{1}{64} = 0.0156 = 1.56\,\% \text{ of the original uranium-235.}$$

Using these data it is easy to calculate that the ^{235}U/^{238}U ratio in natural uranium when the Earth was formed was about 22 %. Using the correct values for the half-lives of the two isotopes the correct value of this percentage is 23.1 %.

Problem 2.4: Activity of antimony. A radioactive isotope ^{124}Sb (antimony), with initial activity $R_0 = 7.4 \times 10^7$ Bq, has a half-life of 60 d. Calculate its residual activity R after one year.

Solution: the decay constant of the sample of antimony is

$$\lambda = \frac{0.693}{60\,\text{d}} = 0.01155\,\text{d}^{-1}.$$

Using Eqs. (2.9) and (2.12) and calling N_0 and N the number of nuclei of Sb at the time zero, and after 1 year (= 365.25 d), respectively, we get:

$$R_0 = \left|\frac{dN}{dt}\right|_0 = \lambda N_0;$$

$$R = \left|\frac{dN}{dt}\right| = \lambda N = \lambda N_0 e^{-\lambda t} = R_0 e^{-\lambda t} = 7.4 \times 10^7 e^{-0.01155 \times 365.25}$$
$$= 1.09 \times 10^6 \text{ Bq.}$$

Problem 2.5: Half-life and decay constant of a radionuclide. The activity of a radioactive isotope decreases from 350 to 275 disintegrations per minute in 3.5 h. Calculate the half-life of the radionuclide and its decay constant.

Solution: From the previous problem we have

$$R/R_0 = e^{-\lambda t}.$$

Extracting the natural logarithm of the first and second member of this equation, one gets:

$$\ln \frac{R}{R_0} = -\lambda t.$$

From which it follows

$$\lambda = -\frac{1}{t} \ln \frac{R}{R_0} = \frac{1}{t} \ln \frac{R_0}{R}.$$

That is, noting that 3.5 h = 12,600 s:

$$\lambda = \frac{1}{12,600} \ln \frac{350}{275} = 1.91 \times 10^{-5}\,\mathrm{s}^{-1}.$$

The half-life is then:

$$\tau_{1/2} = \frac{0.693}{\lambda} = \frac{0.693}{1.91 \times 10^{-5}} = 0.3628 \times 10^5\,\mathrm{s} = 10.08\,\mathrm{h}.$$

2.3 Radioactive Families

A decay chain is a sequence of radioactive decay processes, in which the decay of one nucleus creates a new nucleus that is itself radioactive, and this sequence continues until eventually a stable nucleus is reached. For example, uranium-238 decays into thorium-234, which in turn decays into protactinium-234, and so on until stable lead-206 is produced at the end of the chain. The group of nuclides within a series of decays is called *radioactive family*.

All known natural radionuclides are grouped in three families having as the nucleus at the beginning of the series respectively thorium-232 (^{232}Th, $Z = 90$), uranium-238 (^{238}U, $Z = 92$), and uranium-235 (^{235}U, $Z = 92$). All isotopes of these elements are radioactive and the above three are the heaviest ones with a half-life so long that they still exist in nature.

The mass numbers of every isotope in these chains have values of the type $A = 4k$, $A = (4k + 2)$, and $A = (4k + 3)$, with k integer, respectively. A radioactive family of the type $A = (4k + 1)$ is missing in nature, but it has been produced artificially.[2] This family is known as the *neptunium family* from its longest-lived member ^{237}Np$_{93}$, that has a half-life of 'only' 2.14×10^6 years. Its absence in nature is explained by the fact that all its members have lifetimes much shorter than the age of the solar system;

---

[2] Also the elements beyond uranium and thorium in the three other families have been produced artificially.

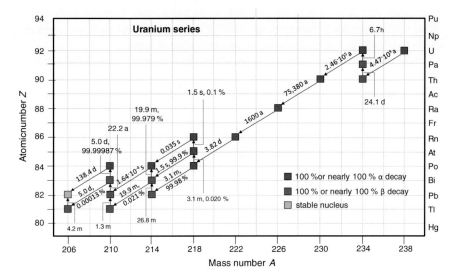

Fig. 2.2 The uranium decay series ($4k + 2$) in the A-Z plane. It is the longest-known decay chain. The half-life of each nucleus is given, together with its decay mode and relative probability (where the latter is not indicated it is equal to 100 %). Values have been rounded off for convenience. Diagonal arrows represent alpha decays, while vertical arrows represent beta decays. Red and blue squares are 100 % or nearly 100 % alpha- and beta-emitting nuclei, respectively. The green square corresponds to the final stable isotope, lead-206

so, even if these radionuclides existed at the time of the formation of the Earth, they are now completely transmuted into the final stable nuclide of the family, which is thallium-205 ($^{205}\text{Tl}_{81}$). Some older sources quote bismuth-209 as the final isotope, but it was recently discovered that it is radioactive, with a half-life of 1.9×10^{19} years. This is the only series that does not end with an isotope of lead. For all the other three families indicated above, the series of subsequent decays continues until it forms a stable isotope of lead ($^{206}\text{Pb}_{82}$, $^{207}\text{Pb}_{82}$ or $^{208}\text{Pb}_{82}$).

Figure 2.2 shows the uranium decay series ($4k + 2$) which has as parent the ^{238}U nucleus and ends with ^{206}Pb. The nuclei are represented by red or blue squares at the corresponding values of Z and A. The diagonal arrows indicate an α-decay (A and Z decrease by 4 and 2 units, respectively), and the vertical ones indicate a β^--decay (A remains the same and Z increases by 1).

The initial nucleus ^{238}U decays by α-emission into ^{234}Th, but this daughter nucleus is itself unstable and undergoes β-decay into the protactinium nucleus, ^{234}Pa. This nucleus is also unstable and decays into ^{234}U, and so on.

For some radionuclides, two alternative decay modes, or *branches*, are possible. It is the case of ^{218}Po in the figure, which can undergo both α- and β-decay leading to the daughter nuclei ^{214}Pb and ^{218}At, respectively. Then, after another decay of the daughter nuclei (β- and α-decay, respectively), both branches lead to ^{214}Bi. Similar cases are those of the isotopes ^{218}At, ^{214}Bi, and ^{210}Bi. When such nuclei decay, they can do so following one path or the other, with different probabilities called *branching ratios*. However, the half-life assigned to those isotopes is just one number referred to a generic decay of one or the other type.

Thanks to these decay chains, it is possible to find in nature radioactive elements with short half-lives that would otherwise not exist. For example, the radium isotope ^{226}Ra$_{88}$ has a half-life so short (1600 years) that all the radionuclides of this species that were produced when the Earth was formed, about 4.5 billion years ago, have now disappeared. However, the decay chain of the uranium-238 family ensures a continuous supply of ^{226}Ra$_{88}$. The same occurs for many other radioactive nuclei.

In a radioactive family, the decay of the parent nucleus acts as a bottleneck for the decay of the daughter nucleus. For example, in comparison to the half-life of 1600 years of the ^{226}Ra$_{88}$, the daughter nucleus radon, ^{222}Rn$_{86}$, decays almost immediately (3.82 days), but cannot decay before being formed. Therefore, it is possible to observe the decay of a ^{222}Rn$_{86}$ nucleus, when one of its parents ^{226}Ra$_{88}$ happens to decay after many years of survival.

2.4 Sequential Decays

If a parent nucleus A decays to a daughter nucleus B that is also unstable, the equation governing growth and depletion of the two species A and B are:

$$\frac{dN_A}{dt} = -\lambda_A N_A, \tag{2.13}$$

$$\frac{dN_B}{dt} = -\lambda_B N_B + \lambda_A N_A, \tag{2.14}$$

where N_A and N_B are the numbers of nuclei of parent and daughter λ_A and λ_B their respective decay constants.

The general solution to this equation for the nucleus B is

$$N_B(t) = N_B(0)e^{-\lambda_B t} + \frac{\lambda_A}{\lambda_B - \lambda_A} N_A(0)(e^{-\lambda_A t} - e^{-\lambda_B t}), \tag{2.15}$$

where $N_A(0)$ and $N_B(0)$ are the numbers of nuclei of parent and daughter at the initial time ($t = 0$). The first term of the right side of this equation represents the number of nuclei B surviving at the time t from the initial number $N_B(0)$, while the second term represents the number of nuclei B produced from the decay of nuclei A and not yet decayed at the time t.

If initially no nuclei B are present, i.e. $N_B(0) = 0$, Eq. (2.15) reduces to

$$N_B(t) = \frac{\lambda_A}{\lambda_B - \lambda_A} N_A(0)(e^{-\lambda_A t} - e^{-\lambda_B t}). \tag{2.16}$$

We note at once that $N_B(t)$ is zero both at $t = 0$ and $t = \infty$, when all nuclei of both A and B have decayed. Clearly, N_B will grow as it is fed by the decay of the parent nucleus, but later as the rate of supply decreases the growth of N_B will be halted and its own decay will become dominant and will set up a decrease in N_B. Then, at some intermediate time t_m the number of nuclei B, N_B, will pass through a maximum value. This is the time t_m for which $dN_B(t)/dt = 0$.

The derivative of Eq. (2.16) with respect to time is zero when

$$\lambda_A e^{-\lambda_A t_m} = \lambda_B e^{-\lambda_B t_m}, \tag{2.17}$$

from which it follows that

$$t_m = \frac{\ln(\lambda_A/\lambda_B)}{(\lambda_A - \lambda_B)}. \tag{2.18}$$

The activity of the nucleus B is $\lambda_B N_B$ (not dN_B/dt). From Eq. (2.16) it follows that

$$\lambda_B N_B = \frac{\lambda_B \lambda_A}{\lambda_B - \lambda_A} N_A(0)(e^{-\lambda_A t} - e^{-\lambda_B t}), \tag{2.19}$$

or, since the activity of A at the time t is $\lambda_A N_A = \lambda_A N_A(0)e^{-\lambda_A t}$,

$$\lambda_B N_B = \lambda_A N_A \frac{\lambda_B}{\lambda_B - \lambda_A}(1 - e^{-(\lambda_B - \lambda_A)t}). \tag{2.20}$$

It is interesting to consider three special cases.

(a) **Daughter nucleus longer-lived than parent**. If $\tau_B > \tau_A$, i.e. $\lambda_B < \lambda_A$, then after a long time $(t \gg 1/\lambda_A)$

$$e^{-\lambda_A t} \ll e^{-\lambda_B t},$$

so Eq. (2.16) may be written

$$N_B = \frac{\lambda_A}{\lambda_A - \lambda_B} N_A(0)e^{-\lambda_B t}. \tag{2.21}$$

Therefore, the amount of nuclei B after a long time is determined only by its own half-life. The initial stock of short-lived nuclei $N_A(0)$ has, in effect, quickly become an initial stock of long-lived nuclei $N_B(0)$ with $N_B(0) \cong N_A(0)$, which decay exponentially like $e^{-\lambda_B t}$. The activity of the daughter nucleus eventually becomes effectively independent of the residual activity of the parent. This case is shown schematically in Fig. 2.3a.

(b) **Daughter nucleus shorter-lived than parent**. If $\tau_B < \tau_A$, i.e. $\lambda_B > \lambda_A$, then after a long time $(t \gg 1/\lambda_B)$

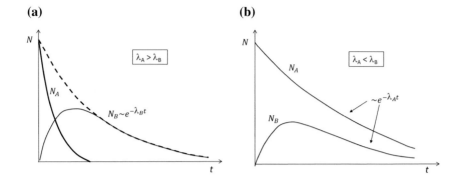

Fig. 2.3 a Decay of a radioactive daughter longer-lived than parent and **b** of a radioactive daughter shorter-lived than parent. In the latter case, at large t, the parent and daughter nuclei decay with the same exponential law, maintaining a constant ratio. The area under a curve in each figure (that is its integral from $t = 0$ to $t = \infty$) represents the total number of disintegrations, hence the total number of radioactive nuclei originally available. In each figure, the curves are not to scale as the areas under the two curves must be equal because the same nuclei are involved, only at different stages in their disintegration series

$$N_B = \frac{\lambda_A}{\lambda_B - \lambda_A} N_A(0)e^{-\lambda_A t}. \tag{2.22}$$

The amount of nuclei B after a long time is determined only by the half-life of the parent nucleus A, as shown in Fig. 2.3b.

From Eq. (2.22) it follows that, for large t:

$$\frac{N_B}{N_A} = \frac{\lambda_A}{\lambda_B - \lambda_A} = \frac{\tau_B}{\tau_A - \tau_B}, \tag{2.23}$$

and recalling that the activity of A and B are $\lambda_A N_A$ and $\lambda_B N_B$ this can be written

$$\frac{\lambda_B N_B}{\lambda_A N_A} = \frac{\lambda_B}{\lambda_B - \lambda_A} = \frac{\tau_A}{\tau_A - \tau_B}, \tag{2.24}$$

meaning that the ratio of the activities of daughter and parent is constant.

This case, where, for large t, the ratio of the activity of daughter to that of parent is greater than unity, is called *transient equilibrium*.

(c) **Daughter nucleus much shorter-lived than parent**. When $\tau_B \ll \tau_A$, i.e. $\lambda_B \gg \lambda_A$, Eq. (2.20) takes on a particularly simple form

$$\lambda_B N_B(t) = \lambda_A N_A(1 - e^{-\lambda_B t}). \tag{2.25}$$

The activity of the daughter nucleus increases according to the simple exponential growth curve governed by its own decay constant, λ_B. In these cases, the equilibrium ratio of the activities becomes substantially unity,

$$\lambda_B N_B = \lambda_A N_A. \tag{2.26}$$

This result can be also found from Eq. (2.24) by letting $\lambda_A \to 0$ in the right-hand side. This condition is called *secular equilibrium*.

If we have a decay chain $A \to B \to C \to D \to \ldots$, with $\tau_B, \tau_C, \tau_D, \ldots \ll \tau_A$, i.e. $\lambda_B, \lambda_C, \lambda_D, \ldots \gg \lambda_A$, we can repeat the same steps and write

$$\frac{dN_A}{dt} = -\lambda_A N_A,$$

$$\frac{dN_C}{dt} = -\lambda_C N_C + \lambda_B N_B = -\lambda_C N_C + \lambda_A N_A,$$

because B is in secular equilibrium with A and we can apply (2.26). It is therefore easy to demonstrate that we also have

$$\lambda_C N_C = \lambda_D N_D = \cdots = \lambda_A N_A,$$

so that all isotopes generated by a decay chain starting with a longer-lived isotope will reach a secular equilibrium where all activities are equal. This is what is found for instance in the decay chain of natural uranium, whose two isotopes 235 and 238 are in secular equilibrium with all 11 and 14 daughter nuclei, respectively.

2.5 Accumulation of Decay Products in a Series of Decays

Let us consider the series of decays

$$A \xrightarrow{\lambda_A} B \xrightarrow{\lambda_B} C \xrightarrow{\lambda_C} \cdots \longrightarrow M \xrightarrow{\lambda_M} N \xrightarrow{\lambda_N} ,$$

that involves n different nuclide-to-nuclide decays. For simplicity, we assume that at time $t = 0$ there are N_0 nuclei of A present and no nuclei of its series of decay products B, C, ..., M, N (that is, $N_B, N_C, \ldots, N_M, N_N = 0$). Let the decay constants of A and its products be $\lambda_A, \lambda_B, \lambda_C, \ldots, \lambda_M, \lambda_N$. Then at a time t, the number N_N of nuclei N present will be given by the integral of

$$\frac{dN_N}{dt} = N_M \lambda_M - N_N \lambda_N, \tag{2.27}$$

where N_M is evaluated from a series of equations similar to Eq. (2.27) for the amounts of the preceding products. This is entirely analogous to Eq. (2.14) for dN_B/dt but the solution requires more mathematical steps, since N_M is a more complicated function than N_B. The result of this integration is

$$N_N = N_0 \left(h_A e^{-\lambda_A t} + h_B e^{-\lambda_B t} + h_C e^{-\lambda_C t} + \cdots + h_M e^{-\lambda_M t} + h_N e^{-\lambda_N t} \right), \qquad (2.28)$$

in which the coefficients are dimensionless functions of the decay constants and have the following systematic values

$$h_A = \frac{\lambda_A}{\lambda_N - \lambda_A} \frac{\lambda_B}{\lambda_B - \lambda_A} \frac{\lambda_C}{\lambda_C - \lambda_A} \cdots \frac{\lambda_M}{\lambda_M - \lambda_A},$$

$$h_B = \frac{\lambda_A}{\lambda_A - \lambda_B} \frac{\lambda_B}{\lambda_N - \lambda_B} \frac{\lambda_C}{\lambda_C - \lambda_B} \cdots \frac{\lambda_M}{\lambda_M - \lambda_B},$$

$$\cdots \qquad\qquad (2.29)$$

$$h_M = \frac{\lambda_A}{\lambda_A - \lambda_M} \frac{\lambda_B}{\lambda_B - \lambda_M} \frac{\lambda_C}{\lambda_C - \lambda_M} \cdots \frac{\lambda_M}{\lambda_N - \lambda_M},$$

$$h_N = \frac{\lambda_A}{\lambda_A - \lambda_N} \frac{\lambda_B}{\lambda_B - \lambda_N} \frac{\lambda_C}{\lambda_C - \lambda_N} \cdots \frac{\lambda_M}{\lambda_M - \lambda_N}.$$

The initial condition, that $N_N = 0$ at $t = 0$, requires that the sum of the coefficients be zero, and these coefficients do satisfy the condition that

$$h_A + h_B + h_C + \cdots + h_M + h_N = 0. \qquad (2.30)$$

The above expressions are valid also if the final nucleus N is stable, in which case it is sufficient to put $\lambda_N = 0$, which immediately implies $h_N = 1$.

The full form of Eq. (2.28), in which the initial concentrations of all daughters nuclei at time $t = 0$ are different from zero (that is, $N_B, N_C, \ldots, N_M, N_N \neq 0$), are called the *Bateman equations* and they appear whenever one wants to describe the full evolution of the nuclides in a decay series.

2.5.1 Approximate Method for Short Accumulation Times

If we consider a time t which is short compared with the mean-life of any member of a series of decays (that is, $t \ll 1/\lambda_i$ for any $i = A, B, C, \ldots N$) approximate solutions can be based on series expansions of the exponential in the general solutions (2.28). In this case, again assuming for simplicity that at $t = 0$ only the A nucleus be present with a number of nuclei N_0, initially for nucleus B we can write

$$dN_B = N_0 \lambda_A dt,$$

($N_0 \lambda_A$ being the activity of A). Therefore, by simple integration, one has

$$N_B = \int_0^t dN_B = \int_0^t N_0 \lambda_A dt = N_0 \lambda_A t, \qquad (2.31)$$

which is equivalent to saying that the initial growth of B is linear with time.

In turn, nucleus C comes from the decay of B, so that we can write

$$dN_C = N_B \lambda_B dt = (N_0 \lambda_A t) \lambda_B dt, \qquad (2.32)$$

which, by simple integration, gives

$$N_C = \int_0^t dN_C = \int_0^t (N_0 \lambda_A t) \lambda_B dt = N_0 \lambda_A \lambda_B \frac{t^2}{2}. \qquad (2.33)$$

We can arrive to the same result by taking the average activity of B in the time from 0 to t, $\langle N_B \lambda_B \rangle t = (N_B \lambda_B / 2) t = (N_0 \lambda_A \lambda_B t / 2)$ and multiplying by t, $N_C = (N_0 \lambda_A \lambda_B t / 2) t = (N_0 \lambda_A \lambda_B t^2 / 2)$. In a similar way we can derive the growth of D, which is given by

$$dN_D = N_C \lambda_C dt = \left(N_0 \lambda_A \lambda_B \frac{t^2}{2} \right) \lambda_C dt, \qquad (2.34)$$

which, by performing one further integration, gives

$$N_D = \int_0^t dN_D = \int_0^t \left(N_0 \lambda_A \lambda_B \frac{t^2}{2} \right) \lambda_C dt = N_0 \lambda_A \lambda_B \lambda_C \frac{t^3}{6}, \qquad (2.35)$$

and so on and so forth. Note that for each nucleus, under these approximations its growth depends only on the decay constants of its precursors, i.e. the growth of B depends only on λ_A, that of C depends only on λ_A and λ_B, that of D depends only on λ_A, λ_B and λ_C, and so on and so forth.

Problem 2.6: Growth of the radioactive daughter. A radioactive source contains only a pure isotope $^{210}\text{Bi}_{83}$. This decays by β^- emission (with half-life $\tau_{1/2A} = 5.0$ d) to $^{210}\text{Po}_{84}$, which in turn decays by α-particle emission (with half-life $\tau_{1/2B} = 138.4$ d) to $^{206}\text{Pb}_{82}$. Calculate at which time T the rate of α–particle emission will reach a maximum.

Solution: The rate of α-decays evolves according to Eq. (2.16). Calling N_A and N_B the numbers of nuclei Bi and Po, and $\lambda_A = \ln 2/\tau_{1/2A}$ and $\lambda_B = \ln 2/\tau_{1/2B}$ the relevant decay constants, the maximum value of α-particle rate is obtained by equating to zero the derivative of that equation.

$$\frac{dN_B}{dt} = \frac{\lambda_A}{\lambda_B - \lambda_A} N_A(0) \frac{d}{dt}(e^{-\lambda_A t} - e^{-\lambda_B t})$$
$$= \frac{\lambda_A}{\lambda_B - \lambda_A} N_A(0)(-\lambda_A e^{-\lambda_A t} + \lambda_B e^{-\lambda_B t}) = 0.$$

This gives

$$-\lambda_A e^{-\lambda_A T} + \lambda_B e^{-\lambda_B T} = 0,$$

from which it follows

$$\frac{\lambda_A}{\lambda_B} = e^{-(\lambda_B - \lambda_A)T},$$

which gives

$$T = \frac{\ln(\lambda_A/\lambda_B)}{\lambda_A - \lambda_B} = \frac{1}{\ln 2} \frac{\tau_{1/2A}\tau_{1/2B}}{\tau_{1/2B} - \tau_{1/2A}} \ln \frac{\tau_{1/2B}}{\tau_{1/2A}} = \frac{1}{0.693} \frac{5.0 \times 138.4}{138.4 - 5} \ln \frac{138.4}{5}$$
$$= 24.9 \,\mathrm{d}.$$

Problem 2.7: Accumulation of a stable end-product in a decay series.
Consider the series disintegration A → B → C → D (stable), with decay
constants λ_A, λ_B, and λ_C. Show that, if A is a very long-lived source, the
number N_D of nuclei D at a time t is given by

$$N_D(t) = N_A(0)\left[1 - e^{-\lambda_A t} + \frac{\lambda_A \lambda_C}{\lambda_B(\lambda_C - \lambda_B)} e^{-\lambda_B t} + \frac{\lambda_A \lambda_B}{\lambda_C(\lambda_B - \lambda_C)} e^{-\lambda_C t}\right].$$

Solution: Using Eqs. (2.28) and (2.29) one has

$$N_D(t) = N_A(0)\left(h_A e^{-\lambda_A t} + h_B e^{-\lambda_B t} + h_C e^{-\lambda_C t} + h_D e^{-\lambda_D t}\right),$$

with

$$h_A = \frac{\lambda_A}{\lambda_D - \lambda_A} \frac{\lambda_B}{\lambda_B - \lambda_A} \frac{\lambda_C}{\lambda_C - \lambda_A},$$

$$h_B = \frac{\lambda_A}{\lambda_A - \lambda_B} \frac{\lambda_B}{\lambda_D - \lambda_B} \frac{\lambda_C}{\lambda_C - \lambda_B},$$

$$h_C = \frac{\lambda_A}{\lambda_A - \lambda_C} \frac{\lambda_B}{\lambda_B - \lambda_C} \frac{\lambda_C}{\lambda_D - \lambda_C},$$

$$h_D = \frac{\lambda_A}{\lambda_A - \lambda_D} \frac{\lambda_B}{\lambda_B - \lambda_D} \frac{\lambda_C}{\lambda_C - \lambda_D}.$$

Observing that $\lambda_D = 0$ these become

$$h_A = -\frac{\lambda_B}{\lambda_B - \lambda_A}\frac{\lambda_C}{\lambda_C - \lambda_A}; \ h_B = -\frac{\lambda_A}{\lambda_A - \lambda_B}\frac{\lambda_C}{\lambda_C - \lambda_B};$$

$$h_C = -\frac{\lambda_A}{\lambda_A - \lambda_C}\frac{\lambda_B}{\lambda_B - \lambda_C}; \ \text{and } h_D = 1.$$

Assuming $\lambda_A \ll \lambda_B$ and $\lambda_A \ll \lambda_C$, one can approximate the above equations

$$h_A = -1; h_B = \frac{\lambda_A}{\lambda_B}\frac{\lambda_C}{\lambda_C - \lambda_B}; h_C = \frac{\lambda_A}{\lambda_C}\frac{\lambda_B}{\lambda_B - \lambda_C}; \ \text{and } h_D = 1.$$

So that one has

$$N_D(t) = N_A(0)\left[1 - e^{-\lambda_A t} + \frac{\lambda_A \lambda_C}{\lambda_B(\lambda_C - \lambda_B)}e^{-\lambda_B t} + \frac{\lambda_A \lambda_B}{\lambda_C(\lambda_B - \lambda_C)}e^{-\lambda_C t}\right].$$

2.6 Penetrating Power of Nuclear Radiation

Radiation from radioactive nuclei produces specific physical effects when it crosses materials. In particular, the interaction between radiation and matter can result in damages to the crossed materials. For instance, electronic circuits can be easily damaged by certain types of radiation. This is also true for the cells of living beings, in which case a health hazard is implied.

The knowledge of the activity of a radioactive material is not sufficient to evaluate entirely the effects produced by the specific radiation. This stems from the fact that the interactions of nuclear radiation with matter are different and depend on the type and energy of the radiation, as schematically shown in Fig. 2.4. By knowing the ability of the different types of radiation to penetrate matter one can gain an understanding on how to design an appropriate shielding in order to protect people, the environment and sensitive equipment.

Alpha particles can travel only a few centimetres in the air and can be stopped by a sheet of paper or a layer of skin. Beta particles can travel metres in the air and several millimetres in the human body; they can be stopped by a sheet of metal of a few millimetre thickness. Gamma rays, X-rays, and neutrons are more penetrating; thick barriers of dense metal like lead or thick walls of concrete best stop gammas, while neutrons are best shielded by thick layers of concrete or by materials rich in hydrogen atoms, such as water or paraffin. Neutron absorbers like boron or cadmium are also used for their ability to capture neutrons. When shielding neutrons, a special consideration must be given to secondary gamma rays that are produced when neutrons scatter off nuclei or are captured by them.

Fig. 2.4 The penetrating
power of different radiations

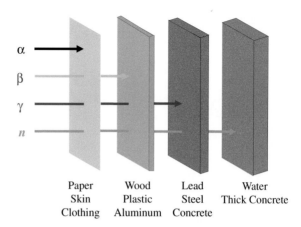

	Paper	Wood	Lead	Water
	Skin	Plastic	Steel	Thick Concrete
	Clothing	Aluminum	Concrete	

Such a different behaviour depends on the various ways of interaction with matter of the various types of radiation.

Alpha and beta particles, which are charged particles, when passing through matter, collide with atoms and molecules encountered along their path. In these collisions, they can either excite atomic levels of atoms and molecules or knock electrons out of them (that is, they can ionise them). In both processes, alpha and beta particles transfer a fraction of their energy to the atomic electrons. Because of this energy loss, they slow down and eventually come to rest. At the end of their path, when they stop, their initial energy has been entirely transferred to the crossed material. Consequently, α- and β-particles have a characteristic average traveling distance (so-called *range*) beyond which they are effectively fully absorbed. This range depends on the energy of the ionising particles and the nature of the crossed material. Typically the energy loss per unit length-of-path of alpha-particles in crossing matter is very high and that of beta-particles is lower than alphas but still significant; this is why relatively thin layers of light material (air, paper, wood, plastic, human skin) are highly effective in stopping them and can therefore perform a shielding function.

Gamma rays and X-rays are both forms of electromagnetic radiation, like light and radio waves, but of much higher frequency (and hence shorter wavelength). They can also be thought of as beams of quantum particles, *photons*, each carrying a discrete quantity of energy but having no mass. Gamma and X-rays have the same basic properties, but differ in their origin. X-rays are emitted by atomic electrons, that is, from processes outside the nucleus, while γ-rays originate inside the nucleus. X-rays are also generally lower in energy and, therefore, less penetrating than γ-rays. Gamma rays and X-rays also ionise the material they pass through, but the way they lose energy is quite different from alpha particles. A photon can cross significant thicknesses of material retaining its entire energy, until it interacts with an atom and knocks out an electron, transferring to it a good fraction of its energy, or being completely absorbed in the process. The extracted electron, in turn, transfers the received energy to the material, with the same mechanism of β-rays.

So passing through matter, gamma and X-rays eventually disappear by being absorbed by the atoms, instead of losing energy by slowing down (as they are forms of electromagnetic radiation, there is only one possible speed for them: the speed of light). Consequently, gamma and X-rays do not have a specific penetration depth, rather their intensity decreases exponentially toward zero as the depth of the crossed medium increases. Instead of the average penetration depth in a medium, for photons one speaks of *attenuation length*, which is the distance over which their intensity is reduced by a factor $e = 2.71828$.

Neutrons are particles which are also very penetrating. On Earth, they can be produced in the spontaneous fission of nuclei, in the induced fission of nuclei in a nuclear reactor, or by energetic cosmic rays that break up nuclei into their proton and neutron constituents. They have null electric charge and, therefore, they interact weakly with matter and do not directly ionise it. Consequently, they are very penetrating, and cannot be slowed down easily.

Neutrons lose speed in collisions with atomic nuclei of the material they pass through. At low neutron energies collisions are predominantly *elastic*, i.e. are processes in which the total kinetic energy of the interacting bodies is conserved. In Sect. 3.7 it will be shown by simple kinematics considerations that the lighter the atomic nucleus, the higher the energy lost in the collision. This results in a progressive loss of their kinetic energy. To stop neutrons one therefore uses thick layers of concrete or materials that contain hydrogen (like water or paraffin), which absorb their energy more effectively. If there is sufficient material for the neutrons to undergo a sufficient number of collisions, this energy loss goes on until neutrons are "thermalized", i.e. reach equilibrium with the thermal energy of atoms in the medium they move. In this condition, they have a mean energy $\sim kT$ (where $k = 1.38 \times 10^{-23}$ JK^{-1} is the Boltzmann's constant) given by the temperature T of the medium; at ordinary temperature, $T = 300\ °C$, $kT \approx 0.025$ eV.

If the incident neutron energy is high enough, an *inelastic* collision can occur. In this case, the recoiling nuclei can be left in an excited state, whose energy is later released by emitting γ-rays, or can be broken up with emission of other particles (photons, neutrons, protons and α-particles). Neutrons can also be captured by certain nuclei, a process that is typically accompanied by the emission of γ-rays, and the isotope formed in the capture process can be unstable and undergo β-decay. In all the above collision processes, emitted gamma rays and charged particles can ionise matter, thereby transferring energy to it.

Like gamma radiation, the range of neutrons is not well defined. The distance they travel depends on the probability of interaction in a particular material. Also for them, one speaks of attenuation length.

In Table 2.3 are given the penetration depth of various types of ionising radiation when entering the human tissue, in increasing order of penetration and for two different kinetic energies of 1 and 10 MeV. For α and β, the penetration depth refers to their range. For neutrons and γ-rays, it is the distance travelled until the number of incoming particles is halved.

For a given specific energy, alpha particles, which have a much greater mass than beta particles (about eight thousand times larger), travel much slower and have

Table 2.3 The penetration depth of various types of ionising radiation in human tissue for two given energies

Energy (MeV)	Penetration depth (mm)			
	α	β	n	γ
1	0.005	5	25	100
10	0.2	50	100	300

more time to interact with the atoms of the material, so they lose energy much more quickly. On top of that, alpha particles have two units of electric charge and therefore the Coulomb force between them and the atoms is twice the one felt by beta particles. As a consequence of both these phenomena, alpha particles are stopped by small material thicknesses, featuring the smallest penetration depth. Neutrons and gamma rays are not hindered by charge and so their penetrating power in our body is large.

By using the same 'interaction-time argument', we understand that higher energy means deeper penetration, hence the difference between the two rows in Table 2.3.

2.7 Dosimetry

The main effect of exposure to radiation of any material—such as human tissue—is the deposition of energy in the material. So the unit used to measure radiation exposure is based on the amount of radiation energy absorbed.

Radiation exposure, usually called *absorbed dose*, is measured in *gray* (Gy), so-called after the British physicist Louis H. Gray (Richmond upon Thames, England, 1905—Hillingdon, England, 1965), a pioneer in the field of X-ray and radium radiation measurement, and their effects on living tissue. One gray is defined as the deposition of 1 J of energy in 1 kg of material; 1 Gy = 1 J/kg.

An old and still used, mainly in the USA, unit of absorbed dose is the *rad*, abbreviation for *radiation absorbed dose*. One rad is equivalent to one hundredth of a gray (1 rad = 0.01 Gy).

To assess the effects of radiation in biological tissues it is not sufficient to know the absorbed dose; one must also take into account the quality of the radiation (some types of radiation are more harmful than others) and the fact that certain tissues are more sensitive than others. For an equal absorbed dose, α-particles are much more potentially harmful than β- and γ-rays because they deposit their energy within shorter distances (see Table 2.3), i.e. into a smaller volume; then the resulting cell damage may be more difficult to repair.

Furthermore, different tissues have different sensitivities to radiation damage. In general, the radiation sensitivity of a tissue is proportional to the rate of proliferation of its cells, and inversely proportional to the degree of cell differentiation. For example, for an equal dose from a given type of radiation, the damage caused to the lungs and reproductive organs is greater than that produced to the bones and teeth.

This also means that a developing embryo is most sensitive to radiation during the early stages of differentiation.

The potential radiation damage also depends on the mode of introduction into the body; for example, α radiation is very dangerous if the emitting radionuclide is inhaled. Instead, for external radiation, the danger is much less, because α-particles can be easily shielded and, indeed, the thickness of the skin is sufficient to stop them.

To account for the quality of radiation, different types of radiation are given different weighting factors, that are used to relate the energy they deposit to the biological significance of the damage they cause; the higher the factor, the greater the damage.

To obtain a quantity that expresses, on a common scale for all ionising radiation, the biological damage to the exposed tissue the *equivalent dose* is used, which is obtained by multiplying the absorbed dose by a dimensionless, weighting factor, w_R, of the radiation considered. Table 2.4 reports the weighting factors w_R of the different types of radiation.

The equivalent dose is measured in *sievert* (symbol Sv), named after the Swedish medical physicist Rolf M. Sievert (Stockholm, Sweden, 1986—Stockholm 1966), renowned for work on radiation dose measurements and research on the biological effects of radiation.

The sievert is defined as the absorbed dose from any radiation that produces the same biological damage as 1 gray of X-rays. Therefore, 1 Sv, unlike 1 Gy, produces the same biological effects regardless of the type of radiation considered. For example, Table 2.4 shows that 1 Gy of beta or gamma radiation has 1 Sv of biological effect; 1 Gy of alpha radiation has 20 Sv effect; 1 Gy of 10–100 keV neutron radiation will cause 10 Sv of biological effect.

Moreover, to account for different sensitivities to radiation damage of the different tissues, one uses the *effective dose*, which is obtained by multiplying the equivalent dose for a dimensionless weighting factor, w_T, of the relevant tissue. Also the effective dose is measured in sievert. Table 2.5 shows the weighting factors w_T of certain tissues and organs.

Table 2.4 Radiation weighting factors w_R of various radiation types

Radiation type and energy range	w_R
X- and γ-rays of any energy	1
Electrons and positrons of any energy	1
Protons of energy >2 MeV	5
α particles, fission fragments, heavy nuclei	20
Neutrons of energy <10 keV	5
10–100 keV	10
100–2000 keV	20
2–20 MeV	10
>20 MeV	5

Table 2.5 Weighting factors of some tissues and organs

Tissue/organ	w_T
Gonads	0.20
Red bone marrow, colon, stomach, lung	0.12
Bladder, breast, liver, oesophagus, thyroid	0.05
Skin, bone surfaces	0.01
Other	0.05

A sievert is a fairly large dose of radiation, so the millisievert and microsievert (abbreviated as mSv and µSv) are often used for the average doses commonly encountered.

An older unit of equivalent dose and effective dose is the *rem* (Roentgen[3] Equivalent in Man), symbol rem. One rem is equal to one hundredth of a Sievert (1 rem = 0.01 Sv, = 10 mSv, or equivalently, 1 Sv = 100 rem).

Both short- and long-lived radioactive materials can present serious hazards, but for somewhat different reasons. Radioisotopes with short half-lives are dangerous for the straightforward reason that they have an extremely high number of decays per second per mass unit. Therefore, they can expose people to very high doses in a short time. On the other hand, the shorter the lifetime the faster the radioactive substance decays to natural levels of radioactivity (see Problem 2.9 and Sect. 2.8) and, therefore, for short-lived radioactive materials a protective shielding is required for a shorter interval of time. Isotopes with mid-to-long half-lives do not expose people as heavily (they have a relatively small number of decays per second per mass unit), but they can keep an entire area significantly radioactive for a very long time, e.g. hundreds or thousands or even tens of thousands of years. That is the main reason why disposing of reactor wastes, which often contain such isotopes, is a non-trivial challenge. At the extreme end are isotopes that are so long-lived that their hazard levels are close to zero. Uranium-238, the isotope left after the fissile uranium-235 isotope is removed, falls into this category with its 4.5 billion years half-life.

Problem 2.8: Residual activity. A hospital has bought a specimen of $^{60}Co_{27}$ for medical irradiation treatments. Calculate the residual activity of the specimen after three half-lives (the half-life of ^{60}Co is about 1925 days).
Solution: According to the definition of half-life, the residual activity of the specimen after $3\tau_{1/2}$ amounts to $1/2^3 = 1/8$ of the initial value.

---

[3]The roentgen was the most commonly used unit of radiation exposure in the 1930s. It was named after German physicist Wilhelm Roentgen (Remscheid, Germany, 1845—Munich, Germany, 1923) the early pioneer of medical X-rays. This unit is obsolete and no longer clearly defined (though as a rule of thumb: 1 roentgen is approximately 10 mSv). The roentgen is essentially a measure of how many ion pairs are formed in a given volume of air when it is exposed to radiation. Therefore, it is not a measure of energy absorbed, or dose.

When the activity of a medical source falls down that much, it may become ineffective for treatment but would still be hazardous in terms of potential exposure and should be disposed of. Disposal is usually performed by special transportation to dedicated repositories.

Problem 2.9: Potential hazard of radioactive materials. As regards the hazard, the most potentially harmful radioactive materials are often neither those with long half-life nor those with short half-life, but those in between. This is because our body would be exposed only to the decays that occur in the course of our life and radionuclides with short half-life decay sufficiently rapidly on a human scale, while those with long half-life produce only very few decays per second.

Evaluate the potential hazard of three beta emitters A, B and C with half-life 3 a, 30 a and 300 a, respectively, for an integrated total exposure of 3 years. Suppose, for simplicity, that the three radioactive substances A, B and C have the same initial number N_0 of radionuclides.

Solution: To estimate the potential danger of the three substances one has to evaluate how many nuclei of the three radioisotopes have decayed after 3 years. Then, using Eq. (2.9), with clear meaning of symbols, one gets that in 3 years

- half of the initial nuclei of the radionuclide A (with half-life 3 a) will have decayed

$$\frac{N_0 - N_A}{N_0} = 1 - e^{-3 \times 0.693/3} = 1 - e^{-0.693} = 1 - \frac{1}{2} = 0.5, \text{ by definition.}$$

- less than 1/10 of the initial nuclei of the radionuclide B (with half-life 30 a) will have decayed

$$\frac{N_0 - N_B}{N_0} = 1 - e^{-3 \times 0.693/30} = 1 - e^{-0.0693} = 1 - 0.933 = 0.067,$$

but the material will remain radioactive for a time interval ten times longer than the substance A.

- less than 1/100 of the initial nuclei of the radionuclides C (with half-life 300 a) will have decayed

$$\frac{N_0 - N_C}{N_0} = 1 - e^{-3 \times 0.693/300} = 1 - e^{-0.00693} = 1 - 0.993 = 0.007;$$

but the material will remain radioactive for a time interval hundred and ten times longer than the substance A and B, respectively. If not suitably shielded, the substance A with the shortest lifetime would be most harmful. On the contrary, if the substance A is well-shielded based on its high activity, but B and C are not, based on their longer lifetime, substance B will pose a higher hazard as almost 1/10 of its nuclei will decay during a 3-year exposure, as opposed to substance C for which less than 1 % of its nuclei will decay during the same time span. On the other hand, substance C will remain radioactive for much longer time, which is also a concern.

2.8 Natural and Artificial Radioactivity

There are many unstable isotopes in nature and their radioactivity is called *natural radioactivity*. In the laboratory, we can produce many other unstable isotopes as a result of various nuclear reactions; we say then that these isotopes are produced artificially and one speaks in these cases of *artificial radioactivity*.

It is worth to point out that from the physical point of view natural and artificial radiation are perfectly equal: the properties of a given isotope are independent of the method by which it was obtained. Therefore, artificial radioactivity is not more dangerous or beneficial than natural radiation.

2.8.1 Natural Radioactivity

Radiation is always a component of the environment and, therefore, a large part of the radiation dose we receive is of natural origin and inevitable. Consequently, life has evolved in, and adapted to an environment that has significant levels of ionising radiation.

For example, in our daily life we are exposed to cosmic rays—mainly originating outside the Solar system and consisting primarily of very high energy protons and atomic nuclei—raining down on Earth. They are thought to be generated through various processes, including the birth and death of stars, and maybe those taking place in the so-called active galactic nuclei. The energies of the primary cosmic rays range from around 1 GeV to as much as 10^{11} GeV. The rate at which these particles arrive at the top of the atmosphere falls off with increasing energy, from about 10,000 per square metre per second at 1 GeV to less than one per square kilometre per century for the highest energy particles. When they arrive at the Earth, cosmic rays collide with the nuclei of atoms in the upper layers of the atmosphere, creating showers of secondary particles, mainly *pions*. The charged pions can swiftly decay to another type of subatomic particle, called *muon*. Unlike pions, these do not interact strongly with matter, and can travel through the atmosphere and penetrate below ground. The rate of muons arriving at the surface of the Earth is such that about one per second passes through a volume the size of a person's head. The Earth's atmosphere and magnetic field act as shield against cosmic radiation, reducing the amount that reaches the Earth's surface. With that in mind, it is easy to understand that the annual dose we get from cosmic radiation depends on what altitude we live at.

Naturally occurring radioactive materials (the so-called *NORM*) are present everywhere: in the soil we walk on and in the buildings we live in, in the food we eat and the water we drink. Radioactive gases are in the air we breathe and even our body is made weakly radioactive by the presence of natural radioactive substances. There is no way on Earth one can get away from natural radiation.

Levels of natural radiation, the so-called *background radiation*, as it is usually called, are not constant everywhere; they vary greatly from one location to another, and at different heights above sea level. For example, people resident in Colorado (USA) are exposed to more natural radiation than residents of the east or west coast of the USA, because Colorado has more cosmic radiation, being located at a higher altitude, and more terrestrial radiation from its soils enriched in naturally occurring uranium. Similarly, people standing in St. Peter's Square in Rome, Italy, are exposed to more natural radiation than people staying in other Roman squares, because of the natural thorium content in the porphyry rocks (the famous "cobblestones"), with which St. Peter's Square is paved.

2.8.2 Artificial Radioactivity

The humankind is not only exposed to natural radiation throughout life, but also to various new sources of radiation created in human activities. They comprise radioactive nuclei produced in the various stages of electricity generation by nuclear fission plants (largely from the reprocessing of the fuel discharged from nuclear plants—so-called *spent fuel*—and to a lesser extent from nuclear fuel manufacturing and energy production, see Chap. 6), or in nuclear weapon tests carried out by a few countries in the 1940s and 1950s, or by irradiation of stable nuclei. The production of the latter is linked to their extended use in medicine, for diagnosis and treatment, and numerous other industrial applications. X-ray generators and particle accelerators used in diagnostic and therapeutic medicine, industry and scientific research are another source of artificial radiation.

Figure 2.5 is a cartoon-view summary of the above.

From all the above it is clear that the amount of radiation we receive depends to some extent also on our way of life, our diet, occupation and use of medical services. For example, dentists administering X-rays to their patients, or miners working underground in certain mines, will receive extra radiation. Airline crews, who spend more time than most people at high altitudes, also receive extra radiation from the effects of cosmic rays. Astronauts in space are subject to a similar exposure. Moreover, the use of X-rays and other medical imaging techniques such as computerised tomography (CT) and positron emission tomography (PET) involve significant exposure to ionising radiation. However, the risks involved in these practices are considered to be outweighed by the benefits to the patient. In some countries, the use of such techniques has grown rapidly in recent years (see Fig. 2.6).

NATURAL RADIOACTIVITY

COSMIC
RADIATION

AND OTHER
TERRESTRIAL
RADIONUCLIDES

RADIONUCLIDES
IN THE FOOD

ARTIFICIAL RADIOACTIVITY

MEDICAL
DIAGNOSTIC

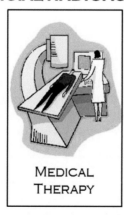

MEDICAL
THERAPY

NUCLEAR
POWER
PLANTS

Fig. 2.5 The humankind has always been subject to natural radiation that is all around us. It comes from the outer space (cosmic radiation), the ground (terrestrial radiation) and even from food. Human beings are also exposed to various artificial sources of radiation. X-ray and other types of radiation used in diagnostic and therapeutic medicine, the radioactive fallout caused by experiments with nuclear weapons or radioactive materials generated during the production of nuclear energy are some examples of these artificial radiation sources

2.9 Average Annual Radiation Dose

As exposure to natural radiation sources is unavoidable, it is interesting to know the average annual effective dose that each individual receives from natural sources of radiation. This is an interesting reference level for every consideration about the dose absorption from man-made radiation sources.

Global (UNSCEAR 2000)

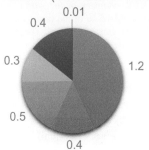

Global (UNSCEAR 2006)

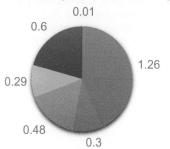

USA (1987)

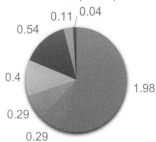

USA (2006)

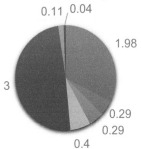

Germany (2005)

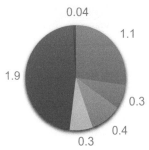

UK (2005)

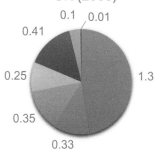

- ■ Radon
- ■ External terrestrial
- ■ Medical

- ■ Cosmic
- ■ Ingestion
- ■ Consumer products

◄ **Fig. 2.6** Estimated contribution in millisievert to public annual exposure from different radiation sources for USA, Germany and United Kingdom and UNSCEAR estimates of the worldwide average annual exposures. For most individuals in the world about 80–85 % of the dose is from natural radiation. Different distributions can be expected for other countries, as all countries considered here have a high level of development. Obtained with data from Ref. [4]

The United Nations Scientific Committee on the Effects of Atomic Radiation (UNSCEAR)[4] has gathered information since 1955 on the typical levels and most important sources of human radiation exposure, and produces a report every few years summarizing the average exposures from all sources. Only occasionally doses to members of the public are directly measured. Usually these doses are assessed on the basis of environmental or effluent[5] monitoring data, using models to simulate environmental exposure scenarios. Moreover, data have different sources of uncertainties reflecting the different way the information is provided by different countries.

The pie charts of Fig. 2.6 show the estimated contribution to public exposure from different radiation sources for different countries, and the UNSCEAR estimates of the worldwide average exposures. Although there is a distinction between natural and man-made radiation, they both affect people in the same way.

Levels of natural background typically range from about 1.5 to 3.5 mSv/a, but are more than 50 mSv/a in several places in Iran, India and Europe and up to 260 mSv/a at Ramsar in Iran, where consequently lifetime doses from natural radiation range up to several thousand millisievert. The highest known level of background radiation affecting a substantial population is in Kerala and Madras States in India where some 140,000 people receive doses which average over 15 mSv per year from gamma radiation, in addition to a similar dose from radon. Comparable levels occur in Brazil and Sudan, with average exposures up to about 40 mSv/a to many people. However, there is no evidence of increased cancer cases or other health problems arising from these high natural levels.

Figure 2.6 shows that the 2006 world natural background radiation exposure is an average of 2.33 mSv/a, while values for some individual countries are: USA 2.96 mSv/a; Germany 2.10 mSv/a, United Kingdom 2.23 mSv/a.

For most individuals, exposure to natural background radiation is the most significant part of their exposure to radiation (79 % for the 2006 world average). Radon[6] is usually the largest natural source of radiation contributing to the exposure of members of the public, sometimes accounting for half the total exposure from all sources.

---

[4]The UNSCEAR was established by the General Assembly of the United Nations in 1955 with the mandate to assess and report levels and effects of exposure to ionising radiation. Governments and organizations throughout the world rely on the Committee's estimates as the scientific basis for evaluating radiation risk and for establishing protective measures.

[5]By effluent here it is meant any gaseous or liquid emission from nuclear installations.

[6]Radon is a radioactive noble gas. Its most relevant isotope is ^{222}Rn, which appears in the uranium decay chain and has a half-life of 3.8 days. Being a heavy gas, it tends to accumulate in underground spaces, such as the basement of houses.

Figure 2.6 also shows that artificial sources of radiation from medical, commercial, and industrial activities contribute about 0.61 mSv/a to the 2006 world annual radiation exposure. The relevant value for the USA is 3.15 mSv/a, Germany 1.94 mSv/a (year 2005), United Kingdom 0.52 mSv/a (2005). Medical exposure, which is by far the largest of these sources of exposure, has grown very rapidly over the last three decades in some industrialised countries. This is clearly shown by the figures for the USA, where the increase in medical uses in the period 1987–2006 resulted in an increase in total annual effective dose per person from 0.54 to 3.0 mSv/a, making medical exposure comparable with the exposure due to natural background radiation. Cleary, the 3.0 mSv/a average is misleading, because the majority of people have only X-rays during their lifetime, whereas a small percentage of people have CT (computed tomography) scans, cancer treatments with radioactive isotopes, angiograms, stent implants, etc. These people have exposures several times greater than 3.0 mSv/a during their treatment periods. At the contrary, the worldwide average increase in the medical exposure is much smaller, increasing only from 0.4 mSv/a in 2000 to 0.6 mSv/a in 2006.

Figure 2.7 gives an estimate of the average radiation dose deposited in the human body by some medical imaging techniques. As it is seen, the radiation dose from a pelvis computed tomography is about three to five times that of the annual naturally occurring background radiation dose; from a chest computed tomography is from twice to more than three times the annual background radiation dose; from a

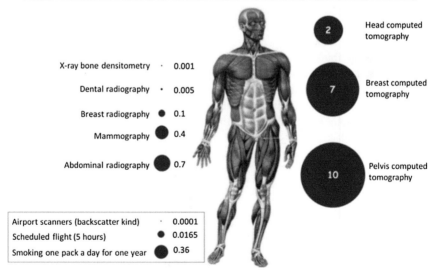

Fig. 2.7 Average radiation doses on the body, in mSv, from various medical imaging techniques. The area of the disks is proportional to the absorbed dose. Also shown are the doses from airport scanners, 5 h of scheduled flight, and smoking one pack a day for one year. Obtained with data from Ref. [5]

mammography is 0.4 mSv, and from a dental radiography is 0.005 mSv. The figure also gives the average absorbed dose from airport X-ray backscatter scanners, five hours of scheduled flight, and smoking one pack of cigarettes a day for one year.

2.10   Biological Effects of Radiation

When one is exposed to radiation, that is when one absorbs some radiation energy, it is said that one gets a *dose of radiation*. This does not make the person radioactive or cause him or her to become contaminated.[7]

Radiation that causes ionisation within human tissues can have very harmful effects. However, like with many other substances we introduce in our body (coffee, alcohol, smoke, prescription drugs, etc.), possible effects of radiation on human health can be evaluated correctly when one knows the intensity of the radiation received and the time-span over which exposure occurs.

For example, one can drink a shot of whiskey without suffering significant side effects. The situation may be different if, instead, one drinks a whole bottle of whiskey. In this case, it is also necessary to know whether it is drunk in a few minutes, in a day or over a longer period of time.

It must also be noted that for many natural substances, threshold values exist below which the substances are beneficial, or even essential, for human life, while they become dangerous above such values. This is, for example, the case of many metals (such as zinc, iron, selenium, and others), vitamins, and also of oxygen itself.

Radiation damage to living tissues results from the transfer of energy to atoms and molecules in the cellular structure, which is the fundamental component of biological tissues. It is composed for 80 % by water (H_2O) and for the remaining 20 % by complex biological structures. Every cell of the body contains molecules of deoxyribonucleic acid (DNA) which store genetic information and control cell growth, functions, development and reproduction. Ionising radiation causes atoms and molecules to become excited or ionised. These excitations and ionisations can cause harmful biological effects, depending on the part of the cell affected. If the ionisation takes place in the DNA, its functions can be altered with serious physical harm up to cancer or genetic modifications.

Our cells have effective repair mechanisms in place against some types of damages, including those induced by radiation (in fact, chromosome aberrations

---

[7]Radioactive contamination is the deposition of, or presence of radioactive substances on surfaces or within solids, liquids or gases (including the human body), where their presence is unintended or undesirable. While radiation exposure and radiation dose are referred to a situation where the person absorbed energy in the form of gamma rays, alpha particles, beta particles or neutrons, contamination occurs when some quantity of radioactive material is transferred on the person's body or on their clothes or shoes. This can happen for instance when the radioactive material is in a form easily transferrable from one place to another, i.e. when it is in the form of a fluid, dust, shavings, airborne particulate, etc.

occur constantly in our bodies). Therefore, injured or damaged cells can repair themselves, resulting in no residual damage. If unable to repair themselves, the cells may die or may perform their functions incorrectly. In particular, they may start to reproduce at an uncontrolled rate, which could lead to cancer.

Millions of body cells die every day, being replaced through normal biological processes. Consequently, cell death due to radiation exposure is an issue only if the number of cells involved becomes so significant as to destroy tissues or impair the functionality of some organs. Broadly speaking, for low doses (such as that received every day from background radiation) damage to the DNA of the cells is the dominant effect, while for high doses cell death prevails.

It is difficult to estimate cancer induction risks, because most of the radiation exposures that humans receive are very close to background levels. At low dose levels of less than 100 mSv spread out over long periods of time, the risk of radiation-induced cancers is so low that it is not readily distinguishable from normal levels of cancer occurrence. In addition, leukaemia or solid tumours induced by radiation are indistinguishable from those that result from other causes. On the contrary, high radiation doses (above 100 mSv in a short period of time, also called *acute doses*) can kill so many cells that tissues and organs are damaged immediately, in what is called *Acute Radiation Syndrome*. The higher the radiation dose, the sooner the effects of radiation will appear, and the higher the probability of death.

A linear, no-threshold (LNT) dose response relationship is used to describe the relationship between radiation dose and the probability for occurrence of cancer. This dose-response hypothesis assumes that there is no threshold, that is to say there is no exposure level below which the risk is zero. It also assumes that any increase in dose, no matter how small, results in an incremental increase in risk: if the exposure doubles, the risk also doubles.

Some scientists strongly dispute the no-threshold assumption. Because life has successfully evolved in the presence of significant levels of natural radiation, they suggest a threshold hypothesis according to which very small exposures are harmless. Other scientists even assert that low levels of radiation are beneficial to health (this idea is known as *hormesis*) [6, 7]. Available data are not clearly enough in favour of any model.

Although generally considered conservative, the LNT model is used worldwide for radiation protection regulations. Using a LNT risk model, the US National Academy of Sciences Committee on the Biological Effects of Ionising Radiation (the BEIR Committee)[8] has estimated that a dose of 0.1 mSv creates an additional risk of death from cancer of approximately one in a million [8].

Clearly, acceptance of a risk is a highly personal matter that requires a certain degree of information to allow for aware decisions. Table 2.6 compares the

---

[8]The National Academy of Sciences (NAS) is a private, non-profit organization of distinguished scholars of the USA. Established in 1863, the NAS is charged with providing independent, objective advice to the nation on matters related to science and technology. In particular, the NAS has been periodically tasked to evaluate the overall state of our knowledge about radiation risk since the first BEIR Committee, which issued its report in July of 1956.

Table 2.6 List of some
activities calculated to have a
one-in-a-million chance of
causing death [6]

Activity	Risk
Smoking 1.4 cigarettes	Lung cancer
Eating 40 tablespoons of peanut butter	Liver cancer
Eating charcoal broiled steaks	Cancer
Spending two days in New York City	Air pollution
Driving 65 km in a car	Accident
Flying 4000 km in a jet	Accident
Canoeing for 6 min	Accident
Receiving a dose of 0.1 mSv of radiation	Cancer

potential risk of radiation to other common activities that also create a risk of a
one-in-a-million chance of death [9].

It is therefore clear how the risk associated to a very low radiation dose
(0.1 mSv) is comparable to other activities in everyday life, although such a radiation dose is already about 10 times more than what we receive in a day from
natural background.

2.11 Applications of Ionising Radiation in Medicine, Research and Industry

Today, ionising radiation is widely used in medicine, research and industry, as well
as for generating electricity. In addition, it has useful applications in agriculture,
archaeology, space exploration, law enforcement, geology (including mining),
environment, conservation and preservation of objects of cultural heritage significance and many others. These beneficial uses of radiation are often less known and
publicized than its potential to do harm.

Below we list some of the many beneficial uses of radiation.

2.11.1 Medical Applications

Modern medicine benefits tremendously from the use of ionising radiations, both
for diagnosis and therapy.

Diagnosis ranges from routine and advanced use of X-rays (for radiographs,
radioscopy radiographs and computed tomography (CT)) to injection of radioisotopes—known as radiopharmaceuticals—for gamma imaging (positron emission
tomography (PET), single photon emission computed tomography (SPECT),
scintigraphy, etc.).

Radiopharmaceuticals are used as "tracers" administered to patients. Wherever a tracer goes in the body, it can be followed by suitable detectors which measure its characteristic radiation. It can be made into a chemical compound, that concentrates in particular organs and biological structures, allowing to determine the precise location, shape and biochemical function of the structures to be mapped out. It is thus possible to identify and perform an early diagnosis of diseases such as cancer and degenerative syndromes of the central nervous system (such as Alzheimer's and Parkinson's disease). The cells, in fact, are equipped with highly specific receptors for certain molecules, which are captured even if present in very low concentrations. In more recent years, the design of radiopharmaceuticals, able to bind specifically and selectively to these receptors, has allowed to study the metabolism of the cell at the molecular level and sometimes even in real time.

Positron emission tomography is an important and widely diffused method in which radioactive tracers are used in medicine [10]. When compounds containing short-lived, positron-emitting isotopes are injected into a patient, they can be followed as they concentrate in particular organs. This is possible because the positrons emitted as they decay very quickly encounter their anti-particles, atomic electrons. When this happens, the electron and positron rapidly annihilate each other and two energetic photons are emitted in precisely opposite directions. By detecting such pairs of photons, with nanosecond timing, the place in the body where the radioactive isotope has decayed can be determined quite accurately. By detecting many such pairs over a period of time, the detailed changes in concentrations of the chemicals can be mapped out. The functioning of living hearts or brains and other organs can be studied and their diseases diagnosed. Brain damage suffered in a stroke, for example, can be mapped out precisely. Positron emission tomography has also become an important method for understanding the normal function of the brain and other organs.

The best-known use of radiation for treating disease is radiotherapy, used against cancer, because it destroys diseased tissue with an efficiency greater than for healthy tissue. The radiation used for cancer treatment may come from a machine outside the body (external-beam radiation therapy), or it may come from radioactive material placed in the body near cancer cells (internal radiation therapy). In the latter case, using the above-described mechanism of the capture of radiopharmaceuticals by the cells, a highly effective approach can be the administration to the patient of compounds containing radionuclides, which radiate preferentially inside cancer cells with limited damage to the surrounding healthy tissue.

An innovative and efficient method to treat tumours is hadrontherapy [10]. It consists in irradiating tumours with protons or light nuclei (alphas, carbon ions). Compared to conventional radiotherapy, it presents two main advantages: a better spatial accuracy in targeting the tumour and a more accurate and efficient irradiation of the tumour. Indeed, while X-rays lose energy slowly and mainly exponentially as they penetrate tissues, charged particles release more energy at the end of their range in matter. This makes it possible to target a well-defined region at a specific depth in the body that can be tuned by adjusting the energy of the incident particle beam, with reduced damage to the surrounding healthy tissue.

Ionising radiation is also routinely used for sterilization of needles, surgical instruments, and dressings—especially those made of non-heat-resistant materials — because they kill bacteria and viruses.

2.11.2 Research Applications

X-ray diffraction is used in the study of molecular tridimensional structures from mineral to biological macromolecules. It is the primary method for determining the molecular conformation of biological macromolecules, particularly proteins and nucleic acids such as deoxyribonucleic acid (DNA) and ribonucleic acid (RNA). It is worth to recall that the double-helical structure of DNA was deduced from X-ray diffraction data.

Neutron diffraction is also used to study properties of molecules in solution or to help refine structures obtained by X-ray images. The methods are often viewed as complementary, as X-rays are sensitive to the spatial distributions of atomic electrons and scatter most strongly off heavy atoms, while neutrons are sensitive to the spatial location of nuclei and scatter strongly even off many light isotopes, including hydrogen and deuterium.

Electron diffraction has been used to determine some protein structures, most notably membrane proteins and viral capsids.

In the last decades *synchrotron radiation*[9] has become an increasingly important tool for the study of biological macromolecules and for research in the fields of art, archaeology, and the conservation of objects of cultural heritage significance, being specifically suitable for micro-non-destructive analyses.

In archaeology and geology, radioactive decay has found interesting applications to determine the age of fossils and other objects, dates of very remote events and measure very large time intervals.

A number of methods using either protons or neutrons have been developed for performing elemental analysis of samples in various fields, from environmental analyses to archaeology and history of art. Among the most important methods are Rutherford back scattering (RBS); proton induced X-ray emission (PIXE); neutron activation[10] analysis (NAA); neutron radiography and prompt gamma activation analysis (PGAA).

---

[9]Synchrotron radiation is called the electromagnetic radiation emitted when charged particles are subject to a circular acceleration, e.g. when their trajectory is bent by a magnetic field.

[10]Activation, i.e. the transformation of a stable isotope of a certain element into a radioactive one, occurs when nuclear collisions due to the accelerated beam impinging on surrounding materials are such that either a proton or a neutron can be stripped off a stable isotope of the struck material. The resulting isotope can be a β^+ or β^- emitter. Activation can also occur when a (slow) neutron is captured by a nucleus. The capture process forms an excited nucleus which quickly decays to its ground state by emitting γ-rays. The newly formed isotope containing one more neutron can be unstable and will typically undergo β^- decay.

In agriculture, research involving the development of new plant species employs radioisotopes to engender genetic mutations.

2.11.3 Industrial Applications

Radioisotopes are widely used throughout industry. Often they are source of penetrating gamma rays used for examining metal structures in order to find otherwise imperceptible defects in metallic castings and welds and to highlight imperceptible possible flaws before they cause dangerous failures (for example, this technique is used in the periodic checks of ball bearings and jet turbine blades). Radiography is also used to check the flow of oil in sealed engines and the rate and way various materials wear out. Well-logging devices use a radioactive source and detection equipment to identify and record formations deep within a well hole for oil, gas, mineral, groundwater, or geological exploration.

Furthermore, radioisotopes are used: for producing new materials with specific chemical-physical characteristics, through transmutation reactions; in the domestic smoke detectors; for sorting scrap metals and analysing alloys; for the treatment of water and sewage sludge of industrial plants or hospitals; to trace and analyse pollutants; to inspect airline luggage and shipping cargo containers for hidden explosives; etc.

Engineers also use gauges containing radioactive substances to measure the thickness of paper products, fluid levels in oil and chemical tanks, and the moisture and density of soils and materials at construction sites.

The agricultural industry makes use of radiation to improve food production, preservation and packaging. Plant seeds, for example, have been exposed to radiation to bring about new and better types of plants. Besides making plants stronger, radiation can be used to control insect populations, thereby decreasing the use of dangerous pesticides. Irradiation with gamma rays is used to increase the shelf life of certain plants and in some types of pest control of foodstuffs. Irradiation of food can stop vegetables or plants from sprouting after they have been harvested. It also kills bacteria and parasites, and controls the ripening of fruits.

Radioactive material is also used in gauges that measure the thickness of eggshells to screen out thin, breakable eggs before they are packaged in egg cartons. In addition, many of our foods are packaged in polyethylene shrink-wrap that has been irradiated so that it can be heated above its usual melting point and wrapped around the foods to provide an airtight protective covering.

Finally, electricity produced by nuclear fission is one of the greatest uses of radiation. As of 31 October 2015, there were 439 commercial nuclear power reactors operable in 30 countries, with over 380,000 MW of total electric capacity [11, 12]. They provide over 11 % of the world's electricity as continuous, reliable base-load power, without carbon dioxide emissions (see Sect. 4.2 for more details).

2.11.4   Radioactive Dating

A radioactive nucleus has a half-life that is independent of the state of chemical combinations of the atom in which it resides. A collection of radioisotopes therefore forms an excellent clock. It is, thus, possible to establish dates which are of considerable interest in archaeology, history and geology.

In archaeology the *radiocarbon dating*, or ^{14}C-dating, is widely used for dating organic materials (bone, wood, textile fibres, seeds, wood charcoal). The method, based on the measurement of the relative abundances of the isotopes of carbon, allows the dating of materials from ages between 50,000 and 100 years ago.

Carbon, which is a chemical element essential for life, is present in all organic substances. In nature there are three carbon isotopes, two are stable (^{12}C and ^{13}C) and one is radioactive (^{14}C). The latter transforms into nitrogen (^{14}N) by β^--decay, with a half-life of 5700 years. Consequently, it would disappear in the long run, if it were not continually replenished. The production of new ^{14}C occurs in nature, in the upper atmosphere, from the capture of neutrons by nitrogen atoms (the neutrons are produced by the cosmic rays that constantly bombard the Earth and smash atmospheric nuclei), via the specific reaction

$$n + {}^{14}N \rightarrow {}^{14}C + p.$$

The dynamic equilibrium that is established between production and radioactive decay keeps the concentration of ^{14}C constant in the atmosphere, where it is presently mainly bonded to oxygen in the form of carbon dioxide. The latter contains the radioactive isotope ^{14}C in the ratio of 1 atomic part in about 10^{12} of the stable isotope ^{12}C.

All living organisms (plants and animals) continually exchange carbon with the atmosphere through the processes of breathing, photosynthesis or feeding. As a result, as long as an organism is alive, all its tissues have the same concentration of ^{14}C as that found in the atmosphere. As soon as the animal or plant dies, its intake of carbon stops, and, from that time, the carbon-14 it contains begins to decrease because it decays without replenishment. Then the concentration of ^{14}C decreases regularly according to Eq. (2.9).

$$C(t) = C_0 \times e^{-\lambda \Delta t}, \tag{2.36}$$

where C_0 is the concentration of carbon-14 in the atmosphere, Δt is the time elapsed since the death of the organism, and λ is the decay constant of ^{14}C.

So by measuring the amount of ^{14}C present in the organic remains, one gets their age according to the following formula:

$$\Delta t = -\frac{1}{\lambda} \ln \frac{C}{C_0}. \tag{2.37}$$

For example, if in an organic specimen the concentration of ^{14}C amounts to a quarter of the natural level, then the specimen is of the age of two half-lives. If it is less than a factor of 1000, then the death of the organism occured more than ten half-lives ago.

This method allows to date the specimens with a margin of error between 2 and 5 % and up to a maximum time of about 50,000 years; for older specimens, the concentration of ^{14}C is too low to be measured with sufficient accuracy. By using special preparations that enhance the ^{14}C content, it is possible to arrive to about 75,000 years ago.

Using radionuclides of longer half-life, one can perform dating further back in time as far as to estimate the age of the solar system (4.5 billion years). For example, the potassium-argon method, based on the β^+-decay of $^{40}K_{19}$ to $^{40}A_{18}$, with a half-life of 1.25 billion years, is reliable in the range between 4.5 billion years and a hundred thousand years. The system rubidium-strontium, based on the β^--decay of $^{87}Rb_{37}$ to $^{87}Sr_{38}$, with a half-life of 48.1 billion years, is reliable in the range between 4.5 billion years and five million years. Both these methods are used in geological dating.

Problem 2.10 Total body dose. A radiological technician works 3.5 h per day at a distance $d = 5.0$ m from a ^{60}Co radioactive source, with 1.0×10^9 Bq activity. The source emits two γ-rays, with energy 1.33 and 1.17 MeV, in fast sequence. Calculate the dose absorbed by the technician per day, knowing that his/her mass is $M = 80$ kg and supposing his/her body has a front section area of 1.5 m^2 and that γ-rays release in the body about 50 % of all their energy.

Solution: The total energy of the two γ-rays released in each decay is $(1.33 + 1.17)$ MeV = 2.50 MeV. Then the total energy per second released by the radioactive source is

$$1.0 \times 10^9 \times 2.50 = 2.5 \times 10^9 \, \text{MeV/s}.$$

This energy is uniformly released in the space around the source. Then, taking into account that the technician works at a distance of 5.0 m from the source, the fraction of the energy intercepted by his body is equal to the ratio of the frontal section area of his body divided by the surface of the sphere of radius d:

$$\frac{1.5}{4\pi d^2} = \frac{1.5}{4 \times 3.14 \times 5^2} = 4.8 \times 10^{-3}.$$

Recalling that γ-rays release only 50 % of all their energy in the technician's body, the energy deposited in the body is:

$$E = \frac{1}{2} 4.8 \times 10^{-3} \times 2.5 \times 10^9 \times 1.6 \times 10^{-13} = 9.6 \times 10^{-7} \, \text{J/s}.$$

Recalling that 1 Gy = 1 J/kg, the dose absorbed by the technician's body per unit time is

$$9.6 \times 10^{-7}/80 = 1.2 \times 10^{-8}\,\text{Gy/s}.$$

Then, in 3.5 h (=12,600 s), the technician absorbs a dose of

$$12{,}600 \times 1.2 \times 10^{-8} = 1.5 \times 10^{-4}\,\text{Gy}.$$

Recalling that the radiation weighting factor of γ-rays is $w_R = 1$, the equivalent dose is 0.15 mSv.

Problem 2.11: Initial and residual activity. A laboratory has 1.49 µg of $^{13}N_7$ with half-life 600 s. Calculate the initial activity of the specimen and its residual activity after 1 h.

Solution: One mole of the radioactive specimen (13 g) contains 6.02×10^{23} nuclei $^{13}N_7$. Therefore, the number of nuclei $^{13}N_7$ initially present in the specimen is:

$$N_0 = \frac{1.49 \times 10^{-6}}{13} 6.02 \times 10^{23} = 6.90 \times 10^{16}\,\text{nuclei}.$$

Using Eq. (2.10) one has:

$$\lambda = \frac{0.693}{\tau_{1/2}} = \frac{0.693}{600\,\text{s}} = 1.16 \times 10^{-3}\,\text{s}^1.$$

Then, using Eq. (2.12), the activity at the time $t = 0$ is:

$$\left(\frac{dN}{dt}\right)_0 = \lambda N_0 = (1.16 \times 10^{-3}\text{s}^{-1})(6.90 \times 10^{16}) = 8.00 \times 10^{13}\,\text{Bq},$$

and the residual activity after 1 h is:

$$\frac{dN}{dt} = \left(\frac{dN}{dt}\right)_0 e^{-\lambda t} = (8.00 \times 10^{13}\,\text{s}^{-1})\,e^{-(1.16 \times 10^{-3}\,\text{s}^{-1})(3600\,\text{s})}$$
$$= 1.23 \times 10^{12}\,\text{Bq}.$$

Problem 2.12: Time-of-flight PET. Positron emission tomography measures the two annihilation photons that are produced back-to-back after positron emission from a radionuclide-tagged tracer molecule, which is chosen to mark a specific function in the body on a biochemistry level. The photons are measured by an annular array of suitable detectors placed around the body of the patient, as shown in the figure.

Consider an annihilation event where the two emitted photons are detected by the detectors A and B positioned on opposite sides of the patient. From the time difference, Δt, of the arrival of the two photons at the detectors A and B (*time-of-flight difference*) it is possible to reconstruct the annihilation position (red star in the figure). Calculate this position on the line connecting the two detectors A and B, knowing that they are at a distance d from each other.

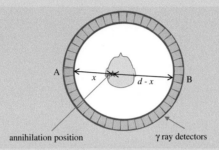

annihilation position   γ ray detectors

Solution: Calling x and $(d - x)$ the distance from the annihilation position and the detectors A and B, respectively; c the speed of the light; and t_A and t_B the times of arrival of the two photons at the detectors A and B, respectively, one can write

$$ct_A = x,$$
$$ct_B = d - x,$$

Subtracting the first equation from the second and calling $\Delta t = (t_B - t_A)$, one gets:

$$c\Delta t = d - 2x,$$

from which it follows that the annihilation occurs at a distance x from detector A:

$$x = \frac{d - c\Delta t}{2}.$$

Time-of-flight PET imaging was investigated in the past, but has never been adopted because of limitations in the available instrumentation. The recent development of scintillators suitable for time-of-flight PET combined with advances in timing resolution and time stability of detector electronics have led to a resurgence of interest in time-of-flight PET scanners, with commercial models being introduced. The advantage of time-of-flight PET, however, remains to be evaluated over the next few years.

Problem 2.13: Radioactive dating. An antique wood relic contains only 6 % of ^{14}C nuclei compared to a specimen of fresh wood. How old is the relic?

Solution: Calling N_0 and N the number of nuclei ^{14}C respectively present in the specimen of fresh wood and in the relic, and T the age of the relic, and using Eq. (2.9), we have:

$$\frac{N}{N_0} = e^{-\lambda T},$$

from which, by taking the logarithm of the first and second member, one gets:

$$\ln \frac{N}{N_0} = -\lambda T,$$

which, using Eq. (2.10) and recalling that the half-life of ^{14}C is $\tau_{1/2} = 5700$ a (see Table 2.2), gives:

$$T = -\frac{1}{\lambda} \ln \frac{N}{N_0} = -\frac{\tau_{1/2}}{0.693} \ln \frac{N}{N_0} = \frac{5700}{0.693} \ln(0.06) \approx 2.3 \times 10^4 \text{ a}.$$

Problem 2.14: Radioactive dating. A sedimentary rock, which includes fossils of prehistoric animals, contains nuclei strontium $^{87}Sr_{38}$ and rubidium $^{87}Rb_{37}$ in the atomic ratio 0.016. Calculate the age T of the fossils assuming the total absence of $^{87}Sr_{38}$ at the time of formation of the rock (the half-life of Rb is 4.81×10^{10} a).

Solution: Calling N_0 the number of initial nuclei Rb, we know that, in the time T, $0.016 \times N_0$ Rb nuclei have undergone β^- decays and formed as many nuclei Sr. Therefore, according to Eqs. (2.9) and (2.10) the age of the fossils is:

$$T = -\frac{\tau_{1/2}}{\ln(2)} \ln(1 - 0.016) = \frac{4.81 \times 10^{10}}{0.693} \times 1.613 \times 10^{-2} = 1.12 \times 10^9 \text{ a}.$$

Problems

2.1 Polonium-210 is an alpha emitter that decays directly to its stable daughter isotope, lead-206, with a half-life of 138.4 days. Evaluate the energy available for this α-decay, knowing that the atomic mass of ^{210}Po is 209.982874 u and that of ^{206}Pb is 205.974465 u.
[*Ans.*: 5.40 MeV]

2.2 Calculate the amount of energy released in the alpha decay of 1 mol of plutonium-239 into uranium-235. (The masses of Pu-239, U-235 and the alpha particle are 239.0521634 u, 235.0439299 u, and 4.002603 u, respectively.)

[*Ans.*: 5.1×10^{11} J]

2.3 The kinetic energy of the α-particle from the decay of ^{226}Ra ($Z = 88$, atomic mass 226.0254098 u) is 4.78 MeV and the recoil energy of the daughter nucleus, ^{222}Rn ($Z = 86$), is 0.09 MeV. Calculate the atomic mass of the ^{222}Rn nucleus.

[*Ans.*: 222.0175785 u]

2.4 Tritium (^{3}H), the heaviest isotope of hydrogen, is radioactive and β$^-$ decays, with half-life of 12.3 years, into a helium isotope ^{3}He. Calculate the maximum value of the energy of the electron emitted in the decay, knowing that the atomic masses of the two isotopes are $M[^3\text{H}] = 3.01604928178$ u and $M[^3\text{He}] = 3.01602932243$ u.

[*Ans.*: 18.6 keV]

2.5 A sample of radioactive material has a half-life of 12 min. If there is 8.0 g of the material at the beginning of an experiment, how much will be left after 1 h has passed?

[*Ans.*: 0.25 g]

2.6 What is the half-life of a radioactive substance whose activity decays to 1/8 of the original value in 150 s?

[*Ans.*: 50 s]

2.7 What percentage of the original nuclei of a sample of radioactive substance are left after: (a) 3 half-lives; (b) 7 half-lives; (c) 10 half-lives; (d) n half-lives.

[*Ans.*: (a) 12.5 %, (b) 0.78 %, (c) 0.098 %, (d) $100(\frac{1}{2})^n$]

2.8 A chemical separation of the lead isotope ^{214}Pb from its radioactive parent is completed at 10:00 a.m. This lead isotope decays with half-life of about 27 min. At 11:00 a.m. there are 8.0×10^6 nuclei of ^{214}Pb remaining. Calculate how many of these nuclei will remain at: (a) 11:27 a.m. and (b) 11.54 a.m.

[*Ans.*: (a) 4.0×10^6 nuclei; (b) 2.0×10^6 nuclei]

2.9 A radioactive material contains 10^{12} radioactive nuclei with half-life of 1 h. Calculate how many of these nuclei decay in 1 s.

[*Ans.*: 1.93×10^8]

2.10 A radioactive antimony isotope ^{124}Sb (half-life 60 days) has an initial activity of 3.7×10^6 Bq. What is its activity after 1 year?

[*Ans.*: 5.5×10^4 Bq]

2.11 A radioactive isotope has an activity of 3.5×10^8 Bq. After 3.90 h, the activity is 2.3×10^8 Bq. What is the half-life of the isotope? What is the activity after an additional 3.90 h?

[*Ans.*: 6.4 h; 1.5×10^8 Bq]

2.12 A sample of a particular β-emitter is placed near a Geiger counter (a detector of β-rays), which registers 480 counts per minute. Twelve hours later the detector counts at a rate of 30 counts per minute. What is the half-life of the material?

[*Ans.*: 3 h]

2.13 In the year 2003, a radioactive sample was observed to decay at a rate of 800 counts per minute. In 2015, that same sample was observed to decay with a rate of 100 counts per minute. What is the half-life of the material in the sample?

[*Ans.*: 4 a]

2.14 Calculate the activity of 1 g of the isotope ^{226}Ra, knowing that its half-life is 1600 a. What will be the activity of the sample after 3200 years?

[*Ans.*: 3.7×10^{10} Bq; 9.2×10^{9} Bq]

2.15 A radiation detector, placed near a sample of the radioactive isotope ^{131}I (half-life 8 days), measures an activity of 32,000 Bq. What was the activity of the sample 40 days before the measurement? How many days must elapse before the activity reduces to 100 Bq?

[*Ans.*: 1,024,000 Bq; 66.6 days]

2.16 A sample of charcoal from an archaeological site contains one-sixteenth of the carbon-14 expected in a fresh sample (the half-life of ^{14}C is 5700 a). Find the age of the charcoal sample.

[*Ans.*: 22,800 years]

2.17 In the bones of a person there are about 175 g of potassium. Calculate the activity of that person, knowing that about 0.01 % of the naturally occurring potassium is the unstable isotope ^{40}K (half-life 1.26×10^{9} a).

[*Ans.*: ≈4590 Bq].

2.18 A chemical analysis of a bone has shown the presence of 300 g di carbon. A measurement of the activity of ^{14}C (half-life 5700 a) has given the value of 10 Bq. Calculate the age of the bone knowing that in a living material there is one ^{14}C atom for every10^{12} ^{12}C atoms.

[*Ans.*: 14,459 years]

References

1. Particle Data Book, Chinese Physics C, **38**, Nb. 9 (2014). Online: http://iopscience.iop.org/cpc
2. Java-based Nuclear Data Information System, http://www.oecd-ne.org/janis/
3. https://www-nds.iaea.org/exfor/endf.htm
4. UNSCEAR, 2008 Report Sources and effects of ionising radiation, Vol. 1 (2008)
5. Le Scienze, Italian edition of Scientific American, **86**, Nb. 513, p. 101, (2011)
6. E.J. Calabrese, *Belle Newsletter* (Biological Effects of Low Level Exposure), **15** Nb. 2, (2009). ISSN 1092

7. Health Physics, vol. 52, issue no. 5 (1987). Monograph volume on Radiation hormesis
8. Health risks from Exposure to low levels of ionising radiation. BEIR VII Report (2006). Online: http://www.nap.edu/
9. DOE Radiation Worker Training based on work by B.L. Cohen, Sc.D. Online: https://www.jlab.org/div_dept/train/rad_guide/effects.html
10. The Nuclear Physics European Collaboration Committee, NuPECC Report "Nuclear Physics for Medicine" (2014). Online http://www.nupecc.org/
11. IAEA Report "Nuclear Power Reactor in the World", Edition 2015
12. IEA, World energy outlook 2014 Factsheet "How will global energy markets evolve to 2040?" http://www.worldenergyoutlook.org/media/weowebsite/2014/141112_WEO_FactSheets.pdf

Chapter 3
Nuclear Reactions and Fission

The first two sections of the chapter present a survey of various types of nuclear collisions and illustrate the cross section concept, which is largely used to describe nuclear reactions. The remaining part of the chapter deals with nuclear fission, the process that is fundamental for the production of nuclear power. Sections 3.3–3.7 examine in detail the distribution of fission fragments, the energy released in the fission reaction, the general conditions for setting up a fission chain reaction, and the way used for slowing down neutrons in thermal neutron reactors. Sections 3.8–3.12 review the basic physical features in fission technology for energy production, and carefully describe the components and the physics of thermal and fast reactors, with a particular attention to reactor control and fuel burnup.

3.1 Nuclear Collisions

A nuclear reaction can be defined as a collision between two nuclei that produces a change in the nuclear composition and/or in the energy state of the interacting nuclei.

Under normal circumstances, nuclear reactions happen when a target is bombarded by particles/nuclei coming from an accelerator or from a radioactive substance. The products of a reaction may include any possible nuclei permitted by the conservation laws (mass-energy conservation, charge conservation, conservation of the number of nucleons, etc.).

When a collision occurs between two nuclei or a particle and a nucleus, several different reactions may take place:

1. One speaks of *scattering* of an incident nucleus/particle colliding with a target nucleus when the incident projectile is found among the products of the reaction.

© Springer International Publishing Switzerland 2016
E. De Sanctis et al., *Energy from Nuclear Fission*, Undergraduate Lecture
Notes in Physics, DOI 10.1007/978-3-319-30651-3_3

The scattering is *elastic* if the interacting nuclei remain intrinsically unchanged in the collision (i.e. the initial and final products are identical) and no other particles are produced. In the process, only a redistribution of the kinetic energy of the colliding nuclei takes place. An example of elastic scattering is the reaction

$$p + {}^{16}O \rightarrow p + {}^{16}O,$$

where a hydrogen nucleus (proton) hits a oxygen nucleus exchanging kinetic energy with it; after the collision the two nuclei change their direction of motion.

The scattering is *inelastic* if the target nucleus (or the projectile) becomes excited, or breaks up. In this case, part of the kinetic energy of the system is transformed into the internal excitation of the colliding nuclei.

An example of inelastic scattering with excitation of the target nucleus is the reaction

$$n + {}^{16}O \rightarrow n + {}^{16}O^*,$$

where the asterisk means that the nucleus ${}^{16}O$ is left in an excited energy configuration (*excited state*).

An example of inelastic scattering with breakup of the target nucleus is the reaction

$$e + {}^{4}He \rightarrow e + {}^{3}H + p,$$

where the target nucleus ${}^{4}He$ breaks into a tritium ${}^{3}H$ and a proton.

2. One speaks of nuclear *transmutation* if there is a rearrangement of nuclear constituents between the colliding nuclei. An example is the reaction

$$d + {}^{14}N \rightarrow {}^{3}He + {}^{13}C,$$

where the nitrogen nucleus loses a proton, transforming into a carbon isotope, and the deuteron (${}^{2}H$) receives a proton becoming a nucleus helium-3.

Other examples are the reactions

$$\alpha + {}^{9}Be \rightarrow {}^{12}C + n,$$
$$p + {}^{23}Na \rightarrow \alpha + {}^{20}Ne.$$

In the first reaction, two protons and one neutron are transferred from the alpha particle (a nucleus ${}^{4}He$) to the beryllium, leaving a free neutron and carbon-12. This reaction is the one in which the neutron was first discovered in 1932 by the English physicist Sir James Chadwick (Bollington, England, 1891—Cambridge, England, 1974. Nobel laureate in physics in 1935) [1].

In the second reaction, a proton and two neutrons of the sodium nucleus (^{23}Na) bind with the incoming proton forming an alpha particle and the remaining twenty nucleons form a neon nucleus ^{20}Ne.

Nuclear reactions may also lead to more than two nuclei in the final state, as it is the case in the reactions

$$p + {}^{14}\text{N} \rightarrow {}^{7}\text{Be} + 2\alpha,$$
$$\gamma + {}^{233}\text{U} \rightarrow {}^{90}\text{Rb} + {}^{141}\text{Cs} + 2n;$$

or just to one nucleus, as in capture reactions, where the incident particle is captured by the target nucleus forming a new system, called a *compound nucleus*. The latter is an unstable structure that undergoes decay or fission into other nuclei. Examples of capture reactions are the following

$$n + {}^{107}\text{Ag} \rightarrow {}^{108}\text{Ag}^{*},$$
$$n + {}^{235}\text{U} \rightarrow {}^{236}\text{U}^{*},$$
$$p + {}^{27}\text{Al} \rightarrow {}^{28}\text{Si}^{*}.$$

where the asterisk as usual indicates an excited state of the nucleus.

Under special circumstances, reactions with more than two reactants are possible. For example, the reaction

$$^{4}\text{He} + {}^{4}\text{He} + {}^{4}\text{He} \rightarrow {}^{12}\text{C},$$

can take place in the overheated plasma of stellar interiors.

Often when a projectile nucleus/particle impinges on a given target nucleus more than one reaction may occur. As an example, the collision of a deuteron with ^{238}U can give rise, among others, to the following reactions:

$$^{2}\text{H} + {}^{238}\text{U} \rightarrow {}^{240}\text{Np} + \gamma,$$
$$^{2}\text{H} + {}^{238}\text{U} \rightarrow {}^{239}\text{Np} + n,$$
$$^{2}\text{H} + {}^{238}\text{U} \rightarrow {}^{239}\text{U} + p,$$
$$^{2}\text{H} + {}^{238}\text{U} \rightarrow {}^{237}\text{U} + {}^{3}\text{H}.$$

In the first of them the deuteron is absorbed by the uranium, forming an excited nucleus of ^{240}Np, that de-excites by emitting a γ-ray. The two following reactions are examples of the so-called *stripping reaction* in which a nucleon (respectively a proton and a neutron in the above examples) is transferred from the deuteron to the target. The last one exemplifies the inverse process: the deuteron captures a neutron from the target and emerges out as tritium. This is denoted as a *pick-up reaction*.

Table 3.1 summarises all the above.

Table 3.1 The main kinds of nuclear collisions

Entrance channel	Exit channel		Examples
$a + X$	$X + a$	Elastic scattering	$p + {}^{12}C \rightarrow p + {}^{12}C$
	$X^* + a$	Inelastic scattering	$p + {}^{12}C \rightarrow p + {}^{12}C^*$
	$Y + b$	Transmutation	$^{7}Li + {}^{208}Pb \rightarrow {}^{197}Au + {}^{18}C$
	$W + c + d + \dots$		$^{7}Li + {}^{208}Pb \rightarrow {}^{212}At + 3n$
	$Y + W + b + \dots$		$^{7}Li + {}^{208}Pb \rightarrow$ two fission fragments + neutrons
	Z^*	Capture reaction	$n + {}^{27}Al \rightarrow {}^{28}Al^*$

3.2 Cross Section

An important quantity for the description and interpretation of a nuclear reaction is the so-called *cross-section*, which quantifies the probability of the occurrence of the reaction. Quantitatively the cross section is related to the nature of the colliding particles and the forces between them and so its measurement gives important information about properties of nuclei and their interactions.

To introduce the concept of cross-section, it is conceptually simplest to consider a reaction with two bodies in both the initial and final states. The following discussion would remain valid also for the case of only one body or more than two bodies in the final state.

Consider the reaction

$$a + b \rightarrow c + d,$$

where a beam of monoenergetic point-like particles a is fired at a thin slice of target material made of particles b, and, as a result of their interaction, the particles c and d are formed (see Fig. 3.1a). The slice should be so thin that a given projectile particle strikes no more than one target nucleus.

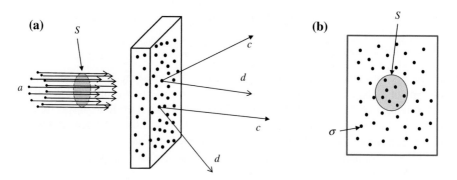

Fig. 3.1 Basic experimental arrangement to measure the cross section of a nuclear reaction. **a** Side view. **b** View along the beam direction

Let n_a be the number of the particles a incident on the target per unit time, and N_b that of the particles b per unit area in the target. We can imagine that the particles a are uniformly distributed over an area S, as shown in Fig. 3.1b, however this is irrelevant for the following discussion as the quantities that we will define are independent of the area S.

Some particles a will pass through the target without interacting, while others will interact with the particles b. Suppose that n_c and n_d particles c and d per time unit are produced, then the cross-section (usually denoted as σ) of the reaction is defined as the ratio

$$\sigma = \frac{n_c}{n_a N_b}. \tag{3.1}$$

Since in the process equal numbers of particle c and d are produced, one can also write

$$\sigma = \frac{n_d}{n_a N_b}. \tag{3.2}$$

Then, the cross section of a reaction is the fraction of the incident particles that undergo the specific interaction, divided by the number of target particles per unit area of a thin target. Clearly, the cross-section multiplied by the incident rate of particles a (n_a particles per unit time) and by the number of target particles N_b per unit area gives the number n_c (n_d) of particles c (and d) produced per unit time.

From Eqs. (3.1) or (3.2) it follows that the dimensions of the cross section σ are those of an area, because both n_a and n_c or n_d have the dimensions of an inverse time and N_b has the dimensions of an inverse square length. So we can pretend that to each target particle b is associated a hypothetical area σ, perpendicular to the incident beam, such that if the projectile particle a crosses this surface σ, there is an interaction and a reaction is produced; while if the particle a misses the surface σ, no reaction occurs. It is important to understand that the cross section σ is a fictitious area which needs not be related to the cross sectional area of the struck target. It is a quantity related to the range of the forces at work in the specific reaction. As such, it is not a true physical property of the target particle, but rather a convenient method for visualizing and comparing probabilities of various types of competing interactions.

For nuclear and particle collisions, a convenient unit for the cross section is the fermi squared ($1 \, \text{fm}^2 = 10^{-30} \, \text{m}^2$). Also frequently used is the barn[1] (symbol b) which is equal to $10^{-28} \, \text{m}^2$.

---

[1]The origin of the barn unit is said to lie in the American colloquialism "big as a barn", which was first applied (in 1942) to the cross sections for the interaction of slow neutrons with certain atomic nuclei, which was much higher than expected.

From the above definition of cross section (Eqs. 3.1 or 3.2) it follows that the probability for any bombarding particle to undergo an interaction process is

$$P = \frac{n_c}{n_a} = \frac{n_d}{n_a} = \sigma N_b, \qquad (3.3)$$

which is the projected total cross section of all target nuclei lying within the unit surface, as seen along the beam direction (Fig. 3.1b).

In general, a given bombarding particle and target can react in a variety of ways, producing a variety of reaction products n_1, n_2, n_3, ... n_m per unit time. Then, in analogy with Eq. (3.1), the total cross section is defined as

$$\sigma_{tot} = \frac{n_1 + n_2 + n_3 + \dots}{n_a N_b} = \frac{\sum_1^m n_i}{n_a N_b}. \qquad (3.4)$$

It is convenient to define also a partial cross section for the i-th process by

$$\sigma_i = \frac{n_i}{n_a N_b}, \qquad (3.5)$$

so that

$$\sigma_{tot} = \sum_i^m \sigma_i. \qquad (3.6)$$

3.2.1   A Convenient Unit for Target Thickness

Usually nuclear collisions are produced bombarding a thin target slab, of thickness dx, with a beam of particles or nuclei: photons, electrons, neutrons, protons, deuterons, α-particles, etc. (as schematically shown in Fig. 3.1). The number N_b of target nuclei per unit area in the target slab is

$$N_b = N_A \frac{\rho dx}{P_A} = N_A \frac{t}{P_A},$$

where ρ and P_A are the density and the atomic weight of the target, respectively, $N_A = 6.022 \times 10^{23}$ is the Avogadro's number, and $t = \rho dx$. The latter is a convenient quantity for calculations because it contains information on both density and thickness of the target. Therefore, it is very often used to express the target thickness instead of the usual dx. Clearly, t has the dimensions of a mass over a square length and is measured in kg/m^2 (or g/cm^2).

Problem 3.1 Rate of collisions. A beam of 1 MeV neutrons, of intensity $I = 5.0 \times 10^8$ cm^{-2} s^{-1} and a cross-sectional area $A = 0.1$ cm^2, hits a thin slab of ^{12}C, which has a thickness $dx = 0.05$ cm. At 1.0 MeV the total cross-section of neutron absorption to ^{12}C is 2.6×10^{-24} cm^2. Calculate at what rate R do interactions take place if the target density is $\rho = 1.6$ g/cm^3. What is the probability that a neutron in the beam will have a collision in the target?

Solution: Calling $M = 12$ the atomic mass of the carbon and $N_A = 6.02 \times 10^{23}$ the Avogadro number, the number density of the target material N_C is

$$N_C = \frac{\rho N_A}{M} = \frac{1.6 \times 6.02 \times 10^{23}}{12} \approx 8 \times 10^{22} \, \text{cm}^{-3}.$$

Then, using Eq. (3.1) gives

$$R = \sigma(IA) \times (N_C dx) = 2.6 \times 10^{-24} \times 5 \times 10^8 \times 0.1 \times 8 \times 10^{22} \times 0.05$$
$$= 5.2 \times 10^5 \, \text{collisions/s}.$$

The probability P of collision of a single neutron is clearly

$$P = \frac{R}{IA} = \frac{5.2 \times 10^5}{5.0 \times 10^8 \times 0.1} = 1.0 \times 10^{-2}.$$

i.e., a single neutron has a 1 % probability of interacting in the target.

Problem 3.2 Production of the radioactive isotope ^{60}Co. A slab of $m = 1$ g of cobalt-59 (^{59}Co) is irradiated by a flux $\Phi = 10^{12}$ s^{-1} cm^{-2} of thermal neutrons. In the reaction, the radioactive isotope ^{60}Co (half-life $\tau_{1/2} = 5.2714$ years) is produced. Calculate the ^{60}Co activity after 1 week of irradiation, knowing that the cross-section for thermal neutron capture on ^{59}Co is $\sigma = 40$ b.

Solution: Calling $A = 59$ the atomic mass of cobalt-59, $N_A = 6.02 \times 10^{23}$ the Avogadro number, and ρ the density of the cobalt slab, the number density of the target material N_{59} is

$$N_{59} = \frac{\rho N_A}{A} = \frac{6.02 \times 10^{23}}{59} \rho \approx 1.0 \times 10^{22} \rho \, \text{cm}^{-3}.$$

Observing that the density of the slab is

$$\rho = \frac{m}{S dx},$$

where S and dx are the surface and the thickness of the cobalt slab, one has:

$$N_{59} = \frac{\rho N_A}{A} = \frac{m}{S dx} \frac{N_A}{A}.$$

Using Eq. (3.1), the production rate of ^{60}Co nuclei is

$$R = \sigma \Phi S N_{59} dx = \sigma \Phi S dx \frac{m N_A}{S A dx} = \sigma \Phi \frac{m N_A}{A}$$

$$= 40 \times 10^{-24} \times 10^{12} \frac{1 \times 6.02 \times 10^{23}}{59} = 4.08 \times 10^{11} \text{ events/s}.$$

Given the long half-life of ^{60}Co, we can neglect the decays occurring during the production week (= 604,800 s) and calculate that the number of ^{60}Co produced in one week of irradiation is

$$N_{60} = 4.08 \times 10^{11} \times 604,800 = 2.47 \times 10^{17}\ ^{60}\text{Co nuclei}.$$

Then, using Eq. (2.12) and observing that 5.2714 years $\approx 1.66 \times 10^8$ s, the activity of ^{60}Co is

$$\frac{dN}{dt} = N_{60} \frac{0.693}{\tau_{1/2}} = 2.47 \times 10^{17} \frac{0.693}{1.66 \times 10^8} = 1.03 \times 10^9 \text{ Bq}.$$

3.3 The Fission Process

Nuclear fission is a reaction in which a nucleus of a heavy element splits into a few lighter fragments (usually two, so-called *binary fission*), releasing a large amount of energy and a certain number of free neutrons (typically two or three). The probability of fission of a nucleus into three parts (*ternary fission*) is from 10^{-2} to 10^{-6} of that of binary fission. Fission into a greater number of parts is of negligible probability at ordinary particle energies.

According to the discussion of Sect. 1.8, fission is energetically possible for a nucleus which has $A > 100$, and, if it does occur, about 200 MeV (=3.2 × 10^{-11} J) of rest mass energy are converted into kinetic energy of the fission fragments and of the released neutrons, and into energy of β electrons, antineutrinos and gamma rays (see Sect. 3.4). The fission products in turn collide with the surrounding atoms and quickly, and within a range of a few to few tens of microns, lose all their kinetic energy, which is converted into thermal energy of the surrounding medium.

It is worth noting that this energy is about 10 million times larger than that released in a chemical reaction, which makes nuclear fission a formidable source of energy, with a very high energy density, i.e. energy produced per unit mass of fuel. This means that, compared to chemical reactions such as combustion of fossil fuels, fission requires a much smaller volume of fuel material to produce an equivalent amount of energy. To make the comparison more quantitative, the energy released by fission of 1 g of ^{235}U (the most widely used nuclear fuel) is about 82 billion joules (see Problem 3.3), equivalent to that released by burning about three million grams of coal (see Problem 3.4).

Although spontaneous fission is energetically possible for a heavy nucleus and is expected to become more probable as the mass number A increases, it is still a rare process even in nuclei as heavy as uranium, being hampered by the intense attractive nuclear forces between nucleons. The nucleus may stretch in an attempt to break down, but generally returns to its equilibrium configuration and vibrates around it. Therefore, the probability of spontaneous fission is very small and the corresponding half-life is very long, about 10^{17} years in the case of ^{235}U [2–4], more than twenty million times the age of the Earth.

In large Z nuclei, fission can be induced by an energy transfer to the nucleus, such as it is obtained by capture of a neutron. In this process, the resulting compound nucleus with one more neutron acquires an abnormal elongated shape and fissions within a very small fraction of a second. For many heavy nuclei, such as uranium-235, the probability of induced fission by neutron capture is very high (half-life of about 10^{-21} s [5]). The capture of a neutron to a nucleus ^{235}U increases its probability of undergoing fission by a factor of approximately 10^{45} (from 10^{17} years to 10^{-21} s). This situation is very favourable for the production of energy. (This process is further described in the Sect. 3.5).

In a fission reaction the emission of free neutrons occurs because heavy nuclei are rich in neutrons. Then, their fission produces very neutron-rich fission products, which are normally some way from the curve of stability of Fig. 1.7. Therefore, they undergo subsequent β^- and γ decays, sometimes with very long half-lives, until they form stable nuclei with the same mass number. An example is the fission reaction[2]

$$^{236}U_{92} \rightarrow {}^{137}I_{53} + {}^{96}Y_{39} + 3n,$$
$$^{137}I_{53} \rightarrow (\beta^- \text{decay}, \tau_{1/2} = 24.5 \text{ s}) \rightarrow {}^{137}Xe_{54} + e + \bar{\nu},$$
$$^{137}Xe_{54} \rightarrow (\beta^- \text{decay}, \tau_{1/2} = 3.818 \text{ min}) \rightarrow {}^{137}Cs_{55} + e + \bar{\nu},$$
$$^{137}Cs_{55} \rightarrow (\beta^- \text{decay}, \tau_{1/2} = 30.08 \text{ a}) \rightarrow {}^{137}Ba_{56} + e + \bar{\nu},$$
$$^{96}Y_{39} \rightarrow (\beta^- \text{decay}, \tau_{1/2} = 5.34 \text{ s}) \rightarrow {}^{96}Zr_{40} + e + \bar{\nu},$$

$$(3.7)$$

---

[2]Often no distinction is made between fission of U-235 and U-236. Also in this book we will speak of fission of U-235 but show reactions involving U-236. Strictly speaking, it is the nucleus U-236 which splits into nuclear fragments, after being formed by capture of a neutron by U-235. However, we also say that the process occurring is the fission of U-235.

or globally

$$^{236}\text{U}_{92} \rightarrow {}^{137}\text{Ba}_{56} + {}^{96}\text{Zr}_{40} + 3n + 4e + 4\bar{\nu}, \qquad (3.8)$$

in which the original nucleus uranium-236 splits into two fragments, iodine-137 ($^{137}\text{I}_{53}$) and yttrium-96 ($^{96}\text{Y}_{39}$) along with three neutrons. The original fragments transform through a series of β^-–decays into the stable barium-137 ($^{137}\text{Ba}_{56}$), and the extremely long-lived zirconium-96 ($^{96}\text{Zr}_{40}$, $\tau_{1/2} = 2.0 \times 10^{19}$ a). In total, three neutrons, four electrons and four antineutrinos are emitted. Note the long half-life of ^{137}Cs ($\tau_{1/2} = 30.08$ a): this is an example of long-lived radioactive waste from a fission reaction (this will be further described in Chap. 6).

Figure 3.2 is a pictorial illustration of the neutron capture reaction by a nucleus ^{235}U, with the formation of compound nucleus ^{236}U, which, then, splits into $^{96}\text{Y}_{39}$ and $^{137}\text{I}_{53}$ emitting three neutrons, as shown in Eq. (3.8).

Fission was discovered in 1939 when the German physicists Otto Hahn (Frankfurt on Main, Germany 1879—Gottingen, Germany 1968. Nobel laureate in Chemistry in 1944) and Fritz Strassmann (Boppard, Germany 1902—Mainz, Germany 1980) discovered the presence of rare-earth elements in uranium after irradiation by neutrons [6]. The Austrian physicists Lise Meitner (Vienna, Austria, 1878—Cambridge, England, 1968) and Otto Robert Frisch (Vienna, Austria, 1904 —Cambridge, England, 1979) then interpreted this production as being due to neutron-induced fission of uranium [7].

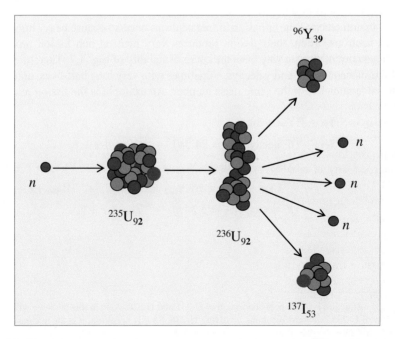

Fig. 3.2 Pictorial representation of a neutron capture reaction by a nucleus $^{235}\text{U}_{92}$ with production of the compound nucleus $^{236}\text{U}_{92}$ that fissions in $^{96}\text{Y}_{39}$ and $^{137}\text{I}_{53}$, thereby emitting three neutrons

3.4   Fission Products

The fission of any particular nucleus can produce many different combinations of fission fragments. For example, Eq. (3.9) lists some of the many possible fission modes of the nucleus $^{236}U_{92}$ that forms with the capture of a neutron by the isotope $^{235}U_{92}$:

$$n + {}^{235}U_{92} \rightarrow {}^{236}U_{92} \rightarrow \begin{cases} {}^{127}Sn_{50} + {}^{105}Mo_{42} + 4n + \sim 178\,\text{MeV}. \\ {}^{136}Te_{52} + {}^{97}Zr_{40} + 3n + \sim 182\,\text{MeV}, \\ {}^{137}I_{53} + {}^{96}Y_{39} + 3n + \sim 179\,\text{MeV}, \\ {}^{139}Ba_{56} + {}^{95}Kr_{36} + 2n + \sim 174\,\text{MeV}, \\ {}^{141}Ba_{56} + {}^{92}Kr_{36} + 3n + \sim 173\,\text{MeV} \\ {}^{144}Ba_{56} + {}^{90}Kr_{36} + 2n + \sim 180\,\text{MeV} \\ {}^{141}Cs_{55} + {}^{93}Rb_{37} + 2n + \sim 180\,\text{MeV}, \\ {}^{139}Xe_{54} + {}^{95}Sr_{38} + 2n + \sim 184\,\text{MeV}, \\ {}^{144}Xe_{54} + {}^{90}Sr_{38} + 2n + \sim 176\,\text{MeV}, \\ {}^{144}La_{57} + {}^{89}Br_{35} + 3n + \sim 168\,\text{MeV}. \end{cases} \quad (3.9)$$

In all the above equations the number of nucleons (protons + neutrons) is conserved (e.g., looking at the first reaction, 1 + 235 = 127 + 105 + 4), but there is a small loss in atomic mass which is released as kinetic energy of fission fragments and neutrons, and as energy of gamma rays (not shown here).

Heavy nuclei (like U-235) which are stable or almost stable have a relatively high ratio of neutrons to protons. This ratio is 1.55 for U-235. Lighter stable nuclei tend to have approximately equal numbers of neutrons and protons. The implication of this difference in neutron to proton ratios for fission is that fission fragments tend to be "neutron rich". If U-235 were split into two approximately equal fragments without the emission of free neutrons then the fragments would have the same neutron to proton ratio as U-235. Thus, fission fragments are generally unstable, having too many neutrons for nuclei of their mass number, and subsequently decay by β^--emission until they form stable nuclei. These decay processes have the net effect of converting a neutron in the nucleus into a proton, thus helping to move the neutron to proton ratio toward stability.

Overall, taking into account also the disintegrations of fission fragments, in multiple fissions of uranium several hundred different radioactive nuclides are produced. Many of these nuclei do not occur naturally on Earth. The reaction in Eq. (3.7) is an example of just one of the many possible fission modes of uranium-236 followed by a series of disintegrations of fission fragments. It is the beta decays, with some associated gamma rays, which makes the fission products highly radioactive (see Chap. 6).

A characteristic feature of fission is that the production of equal mass or near equal mass fission fragments is unlikely and moderately asymmetric fission is the usual outcome. For example, in the fission of uranium-235 the favoured reaction channel is the formation of a lighter nucleus with mass number A between 88 and 103, and a heavier one with A between 132 and 147. A broad spectrum of isotopes is formed with the most common ones being produced in about 6–8 % of fissions, as shown in Fig. 3.3 which represents fission product yields as a function of the mass number A.

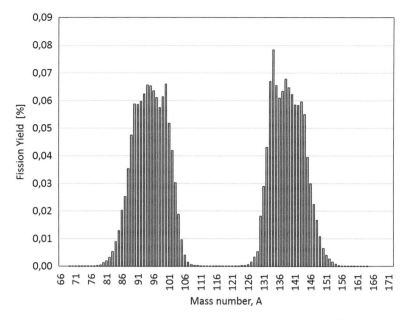

Fig. 3.3 Mass number distributions of fission products from thermal fission of ^{235}U. Obviously, a single mass number corresponds to more than one element. Obtained with data from [8]

(Obviously, a single mass number corresponds to more than one element). In the figure yields are expressed relative to the number of fissioning nuclei, not to the number of fission products, therefore yields sum to 200 % (this simply means that one always has two fragments, neglecting ternary fission that gives rise to three fragments, but is a tiny effect). Similar yield distributions are found for ^{239}Pu and ^{233}U. Since many of the fission products quickly decay, the final composition of the fission products changes with the time elapsed between fission and the measurement. Therefore, if the fragment distribution is measured after some time, only the isotopes with a sufficiently long half-life will have survived.

Fission fragments and their decay products form a significant part of the nuclear waste produced in nuclear reactors. Safety concerns focus on the possibility of their release into the environment. In this respect, the isotopes of greatest concern are those which have high activity and which can cause damage if inhaled or ingested. Among the most important of these isotopes are strontium-90 (^{90}Sr), caesium-137 (^{137}Cs), and several isotopes of iodine. In particular, Sr-90 and Cs-137 have half-lives of around thirty years, so their activity is a concern for hundreds of years (see Sect. 6.3).

3.4.1 Energy Released by Nuclear Fission

From the curve of the binding energy per nucleon *B/A* (see Fig. 1.8) it is easy to calculate that about 200 MeV would be released in the fission of a heavy element.

In fact, the average binding energy per nucleon of a stable nuclide with mass number A between 75 and 160 is approximately 8.5 MeV/nucleon, while that of nuclides with $A \cong 240$ is approximately 7.6 MeV/nucleon. This means that in heavier nuclei the total binding energy can be increased by splitting the original nucleus into two smaller nuclei. Thus, if a nucleus of ^{236}U is divided into two halves having mass number $A = 236/2$, the binding energy per nucleon will increase from $B/A \cong 7.6$ to $B/A \cong 8.5$ MeV/nucleon. This is an increase of about 0.9 MeV/nucleon, or some 210 MeV for a division of a single ^{236}U nucleus.

Taking into account that the atomic mass of ^{235}U is about 218,939 MeV, only about 0.1 % of the total mass of uranium is converted into kinetic energy, the remaining 99.9 % stays in the masses of the protons and neutrons and in the remaining binding energies.

The above value of energy released, which was estimated from the change in B/A, is greater than the actual Q-value of the reaction. This is because the B/A values used in that calculation correspond to stable nuclides, while the hypothetical fission fragments with atomic number $Z/2$ and atomic mass $A/2$ would have too large a neutron excess for stability. The same is true for the actual, asymmetric fission fragments which are typically far from stability and highly excited. This means that the B/A value to be used in the calculation is lower than 8.5 MeV/nucleon, so that the actual energy released is less than 210 MeV. The fission fragments, being highly excited, will release additional energy in several forms, including an average of about 2.5 prompt neutrons per fission, and a cascade of two or three β-disintegrations per each fission fragment. By considering the prompt neutrons in the calculation, the energy release is in the range 170–180 MeV (see Eq. 3.9). If we consider the additional energy released in the form of prompt γ rays and in all subsequent β and γ decays of the unstable fission fragments leading to final stable nuclei, we retrieve the about 200 MeV estimated above.

This qualitative reasoning can be made quantitative using the energy-mass equivalence for any mode of fission. For example in the fission reaction of Eq.Â (3.7) the values of the atomic masses are:

Atomic mass of the compound nucleus	^{236}U	236.045568 u
Atomic mass of	^{137}I	136.917871 u
Atomic mass of	^{96}Y	95.915891 u
Atomic mass of	3 neutrons	3.025995 u
Atomic mass of prompt unstable fragments		235.859757 u
Atomic mass of	^{137}Ba	136.905827 u
Atomic mass of	^{96}Zr	95.908273 u
Atomic mass of	3 neutrons	3.025995 u
Atomic mass of final stable products		235.840095 u

Table 3.2 Energy distribution of fission energy among various fission products

Energy	[MeV]	
Prompt energy of which		177 ± 5.5
Kinetic energy of fission fragments	165 ± 5	
Kinetic energy of prompt neutrons	5 ± 0.5	
Energy of prompt photons	7 ± 1	
Delayed energy from fission product decays of which		23 ± 5.5
Kinetic energy of β-decay electrons	7 ± 1	
Energy of β-decay antineutrinos	10 ± 2	
Energy of γ-decay photons	6 ± 1	
Total		200 ± 6

The mass defect in the whole process (^{236}U $\rightarrow$ ^{137}Ba + ^{96}Zr + $3n$) is $(236.045568 - 235.840095) = 0.205473$ u, which in energy units is 191 MeV. Adding the excitation energy of the parent nucleus ^{236}U, which is in an excited state having absorbed a neutron (see Sect. 3.5), the total energy released in the fission amounts to about 198 MeV. The fraction of energy instantly released at the fission time (that is in the reaction ^{236}U $\rightarrow$ ^{137}I + ^{96}Y + $3n$) is $(236.045568 - 235.859757) = 0.185811$ u = 173 MeV. It is approximately 87 % of the total energy, released as kinetic energy of the fission fragments and the prompt neutrons. The remaining 13 % is emitted partly in the form of prompt γ rays by the primary fission fragments and partly in the subsequent β and γ decays.

Calculations for the other possible fission modes of ^{236}U give similar values, which are also found for the fission of other nuclides (the total energy release varies only by a few MeV between different nuclei). As a practical average value, 200 MeV per fission (=3.20×10^{-11} J per fission) can be used regardless of the fissioning nucleus.

From various measurements, it results that on the average the energy released by fission is partitioned among the various fission products as given in Table 3.2. The largest part of the fission energy is observed as kinetic energy of the fission fragments. It amounts to about 165 MeV on the average. The neutrons emitted have a total average kinetic energy of ~ 5 MeV and the prompt γ-rays emitted in the act of fission account for another ~ 7 MeV. The remaining ~ 23 MeV of fission energy are retained in the fission product nuclei as internal excitation and mass energy. The latter energy is released in a sequence of β-decay steps (accompanied by emission of γ rays) in which the *N/Z* values are adjusted to achieve stability.

In a nuclear reactor, of the 200 MeV, only about 190 MeV are converted into thermal energy, since the antineutrinos escape without heating the crossed medium, as they interact very weakly with matter. Therefore, about 7 % of the thermal energy generated in the reactor originates from radioactive decay of fission products.[3] This decay heat source will remain for some time even after the reactor has been shut down.

Problem 3.3 Energy released by fission of 1 kg uranium-235. Calculate the energy released by the fission of all nuclei $^{235}U_{92}$ contained in a mass of 1 kg. Assume that the energy released by the fission of a single nucleus ^{235}U is about 200 MeV.

Solution: 1 kg of uranium-235 contains

$$N = 6.02 \times 10^{23}/0235 = 2.56 \times 10^{24} \text{ nuclei.}$$

Then, the fission of N nuclei produces an energy:

$$E = 2.56 \times 10^{24} \times 200 \,\text{MeV} = 5.12 \times 10^{26} \,\text{MeV.}$$

Recalling that 1 eV = 4.45×10^{-26} kWh = 1.6×10^{-19} J, this corresponds to:

$$E = (5.12 \times 10^{32} \,\text{eV}) \times (4.45 \times 10^{-26} \,\text{kWh/eV}) = 22.8 \times 10^{6} \,\text{kWh,}$$

and

$$E = (5.12 \times 10^{32} \,\text{eV}) \times (1.6 \times 10^{-19} \,\text{J/eV}) = 8.19 \times 10^{13} \,\text{J.}$$

Using the mass-energy relation, the mass consumed in the transformation is

$$m = E/c^2 = (8.19 \times 10^{13} \,\text{J})/(3 \times 10^8 \,\text{m/s})^2 = 0.9 \times 10^{-3} \,\text{kg} = 0.9 \,\text{g,}$$

that is, only 0.09 % of the initial mass is burned.

Problem 3.4 Energy density of carbon and uranium. The energy released by burning 1 kg of coal is about 2.4×10^7 J. Calculate how many kilograms of coal must be burned to produce the same energy that is released when 1 kg of ^{235}U fissions.

Solution: Using the results of Problem 3.3, the fission of 1 kg of ^{235}U yields 8.19×10^{13} J. Then, the amount of coal that must be burned to produce the same energy is:

$$\frac{8.19 \times 10^{13}\text{J}}{2.4 \times 10^{7}\text{J}} = 3.4 \times 10^{6} \,\text{kg.}$$

---

[3]When the reactor has been operated for a while, an additional heat source is provided by the alpha-decays of transuranic elements produced by neutron capture on ^{238}U.

So, the nuclear energy stored in 1 kg of fissile uranium is equivalent to the chemical energy stored in approximately three million kilograms (or three thousand tonnes) of coal. Fission is, therefore, an extremely powerful source of energy with a very high energy density, i.e. energy produced per unit mass of fuel.

3.5 Fission Induced by Neutron Capture

As said in the previous Section, spontaneous fission of a heavy nucleus is very unlikely and, therefore, it occurs at a very slow pace. However, if one provides the nucleus just with a little additional energy (~ 5 MeV in the heaviest nuclei), fission will occur at a very high rate. This extra energy can be provided in various ways. Neutron capture is an effective one.

The compound nucleus with one more neutron that forms after neutron capture is produced in an excited state, with the excitation energy equal to the sum of the binding energy of the neutron captured in the new nucleus and the original kinetic energy of this neutron. The extra energy increases the agitation of the individual nucleons that causes the compound nucleus to acquire an abnormally-elongated shape with two parts almost separated. The nuclear force between these two parts is weakened, because of their greater separation, and the electric repulsion between protons becomes dominant. Consequently, the nucleus fissions within a very small fraction of a second (about 10^{-21} s).

Fission may take place in any of the heavy nuclei after capture of a neutron. However, low-energy neutrons (so-called *slow* or *thermal neutrons*, see Sect. 3.7) are able to cause fission only in Z-even and N-odd nuclei (that is, nuclei containing an even number of protons and an odd number of neutrons). Such nuclei (such as $^{233}U_{92}$, $^{235}U_{92}$, and $^{239}Pu_{94}$) are said to be *fissile*. For nuclei containing an even number of both protons and neutrons, fission can only occur if the incident neutrons have energy above about 1 MeV (so-called *fast neutrons*). The latter nuclei are called *fissionable*.

Let us analyse in more detail this behaviour by considering the case of uranium that, in order to fission, requires an additional energy of about 5.5 MeV above the ground state.

If the nucleus of the most abundant isotope of uranium found in nature, $^{238}U_{92}$, captures a fast neutron with kinetic energy of 1.0 MeV, the compound nucleus $^{239}U_{92}$ that forms has an excitation energy of 5.8 MeV. It is the sum of the 1.0 meV kinetic energy of the neutron plus 4.8 MeV, which is the binding energy of a neutron in the nucleus $^{239}U_{92}$ (see Problem 3.5). This energy is sufficient for the nucleus to acquire a deformed configuration with more than 5.5 MeV energy above the ground state, and consequently to split in two parts. In other words, fast neutrons with (at least) 1 MeV energy, are effective agents to induce fission of the $^{238}U_{92}$ nuclei, as their kinetic energy makes the process energetically possible.

If, instead, neutrons have a kinetic energy of less than 0.5 MeV (slow neutrons), the compound nucleus $^{239}U_{92}$ that forms has an excitation energy of only 4.8–5.3 MeV. This is not sufficient to produce the large deformation required and fission remains highly unlikely. Consequently, the capture of a slow neutron by the nucleus $^{238}U_{92}$ does not lead to its fission.

The story for capture of neutrons by the rare uranium isotope $^{235}U_{92}$ is different: the binding energy of the captured neutron provides, by itself, sufficient excitation energy to permit the fission of the compound nucleus that forms (see Problem 3.5). In this case, it is not necessary for the incoming neutron to possess a minimum kinetic energy and so, slow neutron capture by $^{235}U_{92}$ can produce fission.

The same happens for the capture of neutrons by the artificial transuranic isotope $^{239}Pu_{94}$, which fissions whatever kinetic energy the captured neutron originally had.

The energy acquired by the nucleus when capturing a neutron can also lead to another process which competes with fission: *radiative capture*. In this case, the extra excitation energy provided by the neutron capture is subsequently released in form of γ rays and no fission occurs. This process can take place whatever the neutron energy both for fissile and fissionable nuclei.

Problem 3.5 Binding energy of the captured neutron to uranium. By neutron capture, the nuclei ^{235}U and ^{238}U transform in the isotopes ^{236}U and ^{239}U, respectively. Calculate the binding energy of the captured neutrons in the two uranium isotopes, knowing that the rest masses of the various isotopes are: $M(U-235) = 235.0439299$ u, $M(U-236) = 236.0455668$ u, $M(U-238) = 238.050784$ and $M(U-239) = 239.054288$ u.

Solution: By repeating the procedure followed in the Problem 1.8 one finds the following mass defects:

$$\Delta m(U-236) = [M(U-235) + M(n) - M(U-236)]$$
$$= 235.043923 \text{ u} + 1.008665 \text{ u} - 236.045562 \text{ u} = 0.007026 \text{ u}.$$

$$\Delta m(U-239) = [M(U-238) + M(n) - M(U-239)]$$
$$= 238.050784 \text{ u} + 1.008665 \text{ u} - 239.054288 \text{ u} = 0.005161 \text{ u}.$$

Then, the binding energy of the last neutron is:

$$\varepsilon(U-236) = (0.007026 \text{ u}) \times (931.494 \text{ MeV/u}) = 6.5446 \text{ MeV},$$

$$\varepsilon(U-239) = (0.005161 \text{ u}) \times (931.494 \text{ MeV/u}) = 4.8074 \text{ MeV}.$$

These energies are found as deformation energies of the isotopes ^{236}U and ^{239}U, respectively.

3.5.1   Uranium Fission Cross Section

Figure 3.4 shows the overall structure of the cross sections for fission of
uranium-235 and uranium-238, by neutron capture (red curves). The abscissa is the
neutron kinetic energy. Notice that both the ordinate and the abscissa have logarithmic scales. Also shown in the figure are the cross sections for the
radiative-capture reaction (n, γ) on both uranium isotopes (green curves).

As shown in the figure, there is no energy threshold for neutron induced fission in
^{235}U, which undergoes fission easily by capturing a slow neutron. The fission cross
section decreases greatly as the neutron velocity increases, until it becomes very
small and approximately constant at neutron energies between 0.5 and 3.0 MeV.

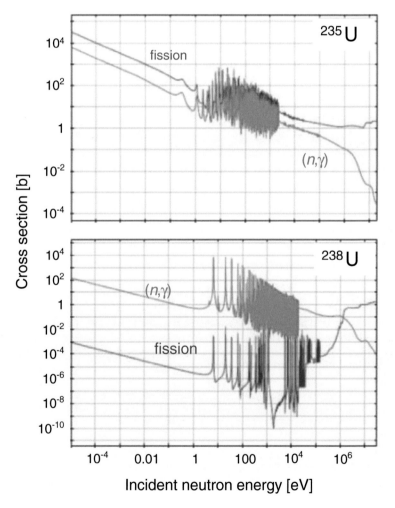

Fig. 3.4 Neutron-induced fission cross sections and radiative-capture cross sections for the
isotopes ^{235}U and ^{238}U, as a function of the incident neutron energy. Note that both scales are
logarithmic. Obtained with data from [9]

Instead, the fission cross section of ^{238}U is very small below about 1.2 MeV, and quickly increases above that energy, which can be considered as a threshold value. However, it is lower than the cross section of ^{235}U in that range of neutron energies. Generally, fissile isotopes (i.e. fissionable isotopes at thermal neutron energies) have a higher fission probability than fissionable isotopes for neutrons of all energies.

The peaks at incident neutron energies between 1 and 10 keV present in both fission and (n, γ) radiative-capture cross sections are called *resonances* and each corresponds to the formation of an excited state of the compound nuclei ^{236}U and ^{239}U.

The radiative-capture cross section for ^{235}U is about a factor of 10 lower than the corresponding fission cross section for slow neutrons. Instead, that for ^{238}U is higher by about a factor of 500 for neutron energies <0.01 MeV. From that value the fission cross section starts to increase, becoming equal to the radiative cross section above about 1 MeV and then higher at higher neutron energy. As we shall see below, these cross sections play an important role in the absorption of neutrons, contributing to reduce the number of neutrons sustaining the chain reaction in a nuclear reactor.

Uranium-238, which makes up about 99.3 % of natural uranium, is not fissionable by slow neutrons, but can produce energy via fast-neutron fission. It has also another important energy application: when exposed to slow neutrons, it can capture one and produce ^{239}U. This nucleus undergoes radioactive β^--decay, with 23.45 min half-life, and forms the transuranic element neptunium-239, ^{239}Np ($Z = 93$), as shown in the following reactions

$$n + {}^{238}\text{U}_{92} \rightarrow {}^{239}\text{U}_{92},$$
$${}^{239}\text{U}_{92} \rightarrow \left(\beta^- \text{decay}, \tau_{1/2} = 23.45 \text{ min}\right) \rightarrow {}^{239}\text{Np}_{93}. \tag{3.10}$$

The nucleus ^{239}Np in turn is radioactive and produces the element plutonium-239, ^{239}Pu ($Z = 94$), by β^- decay with 2.36 d half-life

$$ {}^{239}\text{Np}_{93} \rightarrow \left(\beta^- \text{decay}, \tau_{1/2} = 2.36 \text{ d}\right) \rightarrow {}^{239}\text{Pu}_{94}. \tag{3.11}$$

The importance of this series of reactions is in the fact that plutonium-239 easily undergoes fission induced by thermal neutrons—more easily, in fact, than ^{235}U (the fission cross section of ^{239}Pu is slightly higher than that of ^{235}U).

The plutonium, which is not found in natural ores since it is radioactive with half-life of only 24,110 years, can thus be created by the above process from the abundant uranium isotope ^{238}U. Therefore, both isotopes of uranium can be used in the process of fission induced by slow neutrons: ^{235}U can be used directly and ^{238}U can be converted into ^{239}Pu. We shall see in Sect. 3.11 that the so-called fast breeder reactor, which burns ^{239}Pu and contains ^{238}U, can produce more plutonium than it actually consumes owing to the above series of reactions.

Nuclei that are not fissionable by thermal neutrons, but can be transmuted into fissile material by neutron capture (possibly followed by a radioactive decay) are said *fertile nuclei*.

Besides ^{238}U, also thorium-232 is a fertile nucleus. It transmutes into the fissile isotope ^{233}U through the following steps:

$$n^{232} + \text{Th}_{90} \rightarrow {}^{233}\text{Th}_{90},$$
$$^{233}\text{Th}_{90} \rightarrow \left(\beta^- \text{decay}, \tau_{1/2} = 21.83 \text{ min}\right) \rightarrow {}^{233}\text{Pa}_{91},$$
$$^{233}\text{Pa}_{91} \rightarrow \left(\beta^- \text{decay}, \tau_{1/2} = 26.975 \text{ d}\right) \rightarrow {}^{233}\text{U}_{92}.$$

$$(3.12)$$

The thorium fuel cycle has several potential advantages over a uranium fuel cycle, including thorium's greater abundance (it is roughly three to four times more abundant in the Earth's crust than uranium), better resistance to nuclear weapons proliferation and reduced plutonium and other transuranics production (see Sect. 5.11).

Problem 3.6 Fission cross section of U-235. The fission cross section for thermal neutrons on natural uranium (isotopic composition 0.720 % ^{235}U, 99,275 % ^{238}U, 0.005 % others) is $\sigma = 4.22$ b. Calculate the fission cross section of ^{235}U for thermal neutrons, knowing that, as shown in Fig. 3.4, ^{238}U does not undergo fission with thermal neutrons.

Solution: The fission cross section of natural uranium is the weighted sum of the fission cross sections of the various uranium isotopes, that is:

$$\sigma = \sigma\left({}^{235}\text{U}\right) \times 0.720 \times 10^{-2} + \sigma\left({}^{238}\text{U}\right) \times 99.275 \times 10^{-2} + \sigma(\text{other U}) \times 0.005 \times 10^{-2}.$$

Then, neglecting the last term which has a very low weight, it follows:

$$\sigma\left({}^{235}\text{U}\right) = 4.22/0.0072 = 586 \,\text{b}.$$

Problem 3.7 Yield of radioactive nucleus produced by nuclear bombardment. A gold (^{197}Au) target (thickness $d = 0.01$ cm and density $\rho = 19.3$ g/cm^3) is exposed for a time $T = 10$ min to a constant thermal neutron flux $\Phi = 10^{13}$ s^{-1} cm^{-2}. The neutrons undergo the capture reaction

$$n + {}^{197}\text{Au} \rightarrow {}^{198}\text{Au} + \gamma,$$

with cross section $\sigma = 97.8$ b, forming the nucleus ^{198}Au, which is radioactive and undergoes β-decay (half-life $\tau_{1/2} = 2.7$ d) into ^{198}Hg. Calculate

(a)   the activity of the nucleus ^{198}Au per unit surface in the target at the end of the irradiation;
(b)   the number of nuclei ^{198}Hg per unit surface in the target at the end of the irradiation;
(c)   the maximum number of nuclei ^{198}Au per unit surface in the target (this is the number at equilibrium).

Solution: Let us call, for simplicity, A, B and C the nuclei ^{197}Au, ^{198}Au and ^{198}Hg.

(a) The variation of the population of nucleus B is given by two terms, one representing its growth due to the production by neutron capture by the nucleus A, the other representing its decay

$$\frac{dN_B}{dt} = N_A \Phi \sigma - N_B \lambda_B,$$

where the production term is given by the product of the number N_A of nuclei in the target, times the flux Φ, times the cross section σ, and $\lambda_B = 0.693/(2.7 \times 86,400) = 2.97 \times 10^{-6}$ s^{-1}. By analogy with Eq. (2.14), this equation can be treated mathematically as though there were a parent target A having an activity $\lambda_A = \Phi \sigma = 9.78 times 10^{-10}$ s^{-1} and producing a radioactive substance B. This probability is very small, but the number of target nuclei N_A is very large, hence $\lambda_A N_A$ is non-negligible.

Since $\lambda_B \gg \lambda_A$ (i.e. the ^{198}Au half-life is much shorter than the fictitious one of ^{197}Au) we can use Eq. (2.25) for the activity of the daughter nucleus B

$$N_B(t) = \frac{N_A(t)\lambda_A}{\lambda_B} = \frac{\Phi \sigma}{\lambda_B} N_A (1 - e^{-\lambda_B t}).$$

From which it follows that at the end of the irradiation, i.e. at the time T,

$$N_B(T) = \frac{\Phi \sigma}{\lambda_B} N_A (1 - e^{-\lambda_B T}).$$

Since $\lambda_B T \ll 1$, we can approximate the term within bracket as $\lambda_B T$ (recall that $\lim_{x \to 0} e^x = 1 - x$). In dealing with $x \to 0$, mathematical caution is required. It must be recognized that e^x is not "exactly" equal to unity, because x is not zero but only very small), so that (since $\lambda_A \ll \lambda_B$ we treat N_A as a constant)

$$N_B = \Phi \sigma N_A T = \Phi \sigma \left(N_0 \frac{\rho S d}{P_A} \right) T,$$

where N_0 is the Avogadro number and S and P_A are the surface and atomic mass of the target A. Then, one has

$$\frac{N_B}{S} = \Phi \sigma \left(N_0 \frac{\rho d}{P_A} \right) T = 10^{13} \times 97.8 \times 10^{-24} \times 6.022 \times 10^{23} \frac{19.3 \times 0.01}{197} (10 \times 60)$$

$$= 3.46 \times 10^{14} \text{ nuclei/cm}^2.$$

Then, the activity of the nucleus ^{198}Au per unit surface in the target R_B is

$$R_B = \frac{N_B \lambda_B}{S} = 3.46 \times 10^{14} \times 2.97 \times 10^{-6} = 1.03 \times 10^9 \, \text{Bq/cm}^2$$

(b) The production of mercury nuclei is given by

$$\frac{dN_C}{dt} = \lambda_B N_B(t) = \Phi \sigma N_A (1 - e^{-\lambda_B t}).$$

By integrating this equation from $t = 0$ to $t = T$ one gets

$$N_C(T) = N_A \Phi \sigma T + \frac{N_A \Phi \sigma}{\lambda_B} e^{-\lambda_B T} - \frac{N_A \Phi \sigma}{\lambda_B}.$$

Since $\lambda_B T \ll 1$, one can use the Taylor exponential expansion to second order ($e^{-x} = 1 - x + \frac{1}{2}x^2$), which is a very accurate approximation, so that one gets

$$\begin{aligned}
\frac{N_C(T)}{S} &= \frac{1}{2} \frac{N_A \Phi \sigma}{S} \lambda_B T^2 \\
&= \frac{1}{2} \times 6.022 \times 10^{23} \times \frac{19.3}{197} \times 0.01 \times 10^{13} \times 97.8 \\
&\quad \times 10^{-24} \times 2.97 \times 10^{-6} \times 600^2 = 3.08 \times 10^{11} \, \text{nuclei/cm}^2.
\end{aligned}$$

(c) At equilibrium, the number of B nuclei in the target remains constant, that is the rate of variation (production + decay) is equal to zero: $dN_B/dt = 0$. According to the first equation in (a) describing the growth of nucleus B, at equilibrium $N_A \Phi \sigma = N_B \lambda_B$, and, by using the result on N_B/S obtained in (a), one gets

$$\left(\frac{N_B}{S} \right)_{max} = \frac{N_A \Phi \sigma}{S \lambda_B} = \frac{3.46 \times 10^{14}}{10 \times 60} \frac{1}{2.97 \times 10^{-6}} = 1.94 \times 10^{17} \, \text{nuclei/cm}^2.$$

Problem 3.8 Plutonium production. When U-238 is irradiated by neutrons in a reactor, Pu-239 is produced through the reactions described by Eqs. (3.10) and (3.11): the ^{239}U nucleus that forms following the neutron capture decays into the ^{239}Np nucleus with a half-life of 23.45 min. The latter, in turn, decays to ^{239}Pu with a half-life of 2.36 days.

Calculate the fraction of ^{239}U that transforms into ^{239}Pu in one year, knowing that the flux of neutrons in the reactor (defined as the number of neutrons impinging on both sides of a unit surface per unit time) is $\Phi = 10^{14}$ cm^{-2} s^{-1}, and the radiative capture cross section for thermal neutrons is $\sigma_{cap} = 2.7$ b. Neglect capture reactions on nuclei ^{239}U, ^{239}Np and ^{239}Pu, and the fission of Pu and its decay (which makes sense, as the half-life of ^{239}Pu is much longer than that of ^{239}Np).

Solution: Let us call, for simplicity, A, B, C, D the nuclei ^{238}U, ^{239}U, ^{239}Np and ^{239}Pu, respectively. The rate of capture on A, i.e. the rate of production of B, is clearly given by

$$\frac{dN_A}{dt} = -\sigma_{cap}\Phi N_A = -\lambda_A \Phi,$$

where $\lambda_A = \sigma_{cap}\Phi = 2.7 \times 10^{-10}$ s^{-1}, and N_A is the number of nuclei A. Since the capture reaction makes the nuclei A disappear and the disappearance rate is proportional to the number of nuclei, this equation is mathematically the same as Eq. (2.13) for the decay of nucleus A, even though the physical process is not decay but capture.

The production of B is followed by its decay to C (^{239}Np), which in turn decays to D (^{239}Pu). At $t = 0$, there are only N_0 nuclei A, and zero nuclei B, C, D. Then, the problem is similar to Problem 2.7, and we can write

$$N_D(t) = N_0\left[1 - e^{-\lambda_A t} + \frac{\lambda_A \lambda_C}{\lambda_B(\lambda_C - \lambda_B)}e^{-\lambda_B t} + \frac{\lambda_A \lambda_B}{\lambda_C(\lambda_B - \lambda_C)}e^{-\lambda_C t}\right].$$

In our case, we have $\lambda_A = 2.7 \times 10^{-10}$ s^{-1}, $\lambda_B = 4.9 \times 10^{-4}$ s^{-1}, $\lambda_C = 3.4 \times 10^{-6}$ s^{-1}. Therefore $\lambda_A \ll \lambda_C \ll \lambda_B$, so we can do some approximations and write

$$N_D(t) = N_0\left[1 - e^{-\lambda_A t} + \frac{\lambda_A \lambda_C}{\lambda_B(-\lambda_B)}e^{-\lambda_B t} + \frac{\lambda_A \lambda_B}{\lambda_C \lambda_B}e^{-\lambda_C t}\right],$$

or

$$\frac{N_D(t)}{N_0} = \left(1 - e^{-\lambda_A t} - \frac{\lambda_A \lambda_C}{\lambda_B^2}e^{-\lambda_B t} + \frac{\lambda_A}{\lambda_C}e^{-\lambda_C t}\right).$$

For $t = 1$ year $= 3.15 \times 10^7$ s, only the first exponential survives and we are left with

$$\frac{N_D(t)}{N_A(0)} = \left(1 - e^{-\lambda_A t}\right) = \left(1 - e^{-2.7\times 10^{-10}\times 3.15\times 10^7}\right) = 0.0085 = 0.85\,\%.$$

This means that, starting from one tonne of ^{238}U, after one year 8.5 kg of ^{239}Pu will have been produced.

Problem 3.9 Short-term accumulation of isotopes from decay of fission fragments in a reactor. Consider the following decay chain initiated by the fission process in a nuclear reactor

$$\overset{(n,\,f)}{^{235}\text{U} \longrightarrow {}^{140}\text{Xe} \overset{\tau_{1/2}=16\,\text{s}}{\longrightarrow} {}^{140}\text{Cs} \overset{40\,\text{s}}{\longrightarrow} {}^{140}\text{Ba} \overset{300\,\text{h}}{\longrightarrow} {}^{140}\text{La} \overset{40\,\text{h}}{\longrightarrow} {}^{140}\text{Ce (stable)}}$$

Knowing that the flux of neutrons in the reactor is $\Phi = 10^{14}$ cm^{-2} s^{-1} and the fission cross section for thermal neutrons is $\sigma_f = 590$ b, calculate the accumulation of barium, lanthanum and cerium after 3 h of reactor operation, as a percentage with respect to the initial number of uranium nuclei N_0.

Solution: As seen in the previous two problems, fission induced by neutron capture can be treated mathematically as though there were a parent target ^{235}U decaying into ^{140}Xe with an activity $\lambda_U = \Phi\sigma_f = 5.9 \times 10^{-8}$ s^{-1}. The fission half-life ($0.693/5.9 \times 10^{-8} = 1.17 \times 10^7$ s) is very much longer than the half-lives of ^{140}Xe and ^{140}Cs. The time interval of 3 h is long compared with the half-lives of ^{140}Xe and ^{140}Cs. Then, we can assume that both xenon and caesium will be in secular equilibrium with uranium, so that their activities will be equal to the uranium fission rate without appreciable error. (Analytically, in the general Eq. (2.28), the exponentials which contain the large decay constants of ^{140}Xe and ^{140}Cs become negligible.) On the other hand, 3 h is a short time with respect to the decay times of ^{140}Ba and ^{140}La. Then, by applying Eqs. (2.31), (2.33) and (2.35), the percentage accumulation of Ba in a time t of the order of 3 h will be

$$\frac{N_{\text{Ba}}}{N_0} = \lambda_U t = 5.9 \times 10^{-8} \times 3 \times 3{,}600 \cong 6.4 \times 10^{-4} = 0.064\,\%.$$

The percentage accumulation of lanthanum will be

$$\frac{N_{\text{La}}}{N_0} = \lambda_U \lambda_{\text{Ba}} \frac{t^2}{2} = 5.9 \times 10^{-8} \times \frac{0.693}{300 \times 3{,}600} \times \frac{(3 \times 3{,}600)^2}{2} \cong 2.2 \times 10^{-6}$$
$$= 0.00022\,\%.$$

Finally, the percentage of stable cerium will be

$$\frac{N_{\text{Ce}}}{N_0} = \lambda_U \lambda_{\text{Ba}} \lambda_{\text{La}} \frac{t^3}{6} = 5.9 \times 10^{-8} \frac{0.693}{300 \times 3{,}600} \times \frac{0.693}{40 \times 3{,}600} \frac{(3 \times 3{,}600)^3}{6}$$
$$\cong 3.8 \times 10^{-8} = 3.8. \times 10^{-6}\,\%.$$

3.6 The Chain Reaction

In a fission process, besides the two fragments of medium atomic weight, a few free neutrons are released, depending on the nuclei and on the excitation energy. For example, fission of ^{235}U typically releases 2 or 3 neutrons, with an average of about 2.5. The average number of neutrons emitted upon fission is usually indicated with the Greek letter v.

Some of these neutrons are released immediately, with a delay of less than 10^{-14} s, and are called *prompt neutrons*. A small proportion are released at a later time, and are called *delayed neutrons*. They originate from the decay of neutron-rich fission fragments, or their daughters, which have a half-life of the order of seconds.[4] Although the delayed neutrons are far fewer (only 0.65 % in the case of uranium-235, 0.26 % for uranium-233 and 0.21 % for plutonium-239), they play a crucial role in making a nuclear reactor controllable and in allowing to keep it in the so-called *critical state*, a stable condition where the nuclear chain reaction does not diverge (see below and also Sect. 3.10).

The emission of neutrons in a fission reaction has a great practical importance, because, under appropriate conditions, makes a self-sustaining chain reaction possible. In fact, it is easy to understand that if each of the neutrons resulting from fissions produces other fissions, the number of the latter will grow rapidly and it will suffice only one initial neutron to cause the fission of a huge number of nuclei. This is the keystone of all the practical applications of nuclear energy.

In a nuclear reactor, the fuel and other materials are arranged in a suitable way to produce a self-sustaining chain reaction, where on the average just one of the neutrons released by each fission goes on to cause a further fission. At that point, the reactor is said to have reached *criticality* or is said to be *critical*. In this situation, the energy released per unit time (i.e. the power) is maintained at a constant level and one speaks of *controlled chain reaction*. The minimum amount of fissionable material for a given set of conditions needed to maintain a chain reaction is called *critical mass*.

In the case in which the number of neutrons producing new fissions is smaller than one (*subcritical reaction*), the chain reaction is not self-sustaining and, in the absence of appropriate corrections, dies away. If, on the contrary, the number of neutrons giving rise to new fissions is greater than one, the reactor is said to be *supercritical*: in this case, the energy released per unit time increases and, if left uncontrolled, could lead to a partial or total meltdown of the fissile mass.[5]

---

[4]The fragment components producing the majority of the delayed neutrons have half-life 1.52 and 4.51 s; there are also components with half-life 0.43 s, 15.6 s and 22.5 s.

[5]To this respect, it is important to notice that nuclear reactors cannot explode like an atomic bomb. The latter has all of its fissile material set up for the chain reaction to propagate as quickly as possible. Instead, nuclear power plants use extremely small amounts of nuclear fuel at a time, and the fuel is insufficiently concentrated, so that it cannot explode. This will be discussed more extensively in Chap. 5.

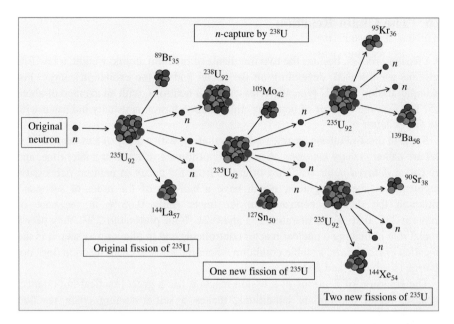

Fig. 3.5 Pictorial representation of a hypothetical chain reaction originated by the fission of a nucleus ^{235}U, induced by neutron capture, into ^{89}Br$_{35}$, ^{144}La$_{57}$ and the release of three neutrons. In the second stage, one of these neutrons is lost, another is absorbed by a nucleus ^{238}U and the third produces another fission of a nucleus ^{235}U into ^{105}Mo$_{42}$ and ^{127}Sn$_{50}$, accompanied by the release of four neutrons. In the next stage, two of these neutrons are lost, and the other two induce fission of two ^{235}U nuclei, respectively into ^{95}Kr$_{36}$ and ^{139}Ba$_{56}$ and into ^{90}Sr$_{38}$ and ^{144}Xe$_{54}$, both accompanied by the release of two neutrons

Figure 3.5 shows a pictorial representation of a hypothetical chain reaction originated by fission of a nucleus ^{235}U, induced by neutron capture, into bromine-89 (^{89}Br$_{35}$) and lanthanum-144 (^{144}La$_{57}$), plus three neutrons. In the second stage, one of these neutrons is lost, another is absorbed by a nucleus of ^{238}U and the third produces a new fission of a nucleus of ^{235}U into molybdenum-105 (^{105}Mo$_{42}$) and tin-127 (^{127}Sn$_{50}$), accompanied by the release of four neutrons. In the next stage, two of these neutrons are lost, and the other two induce fission of two ^{235}U nuclei, respectively into krypton-95 (^{95}Kr$_{36}$) and barium-139 (^{139}Ba$_{56}$), and into strontium-90 (^{90}Sr$_{38}$) and xenon-144 (^{144}Xe$_{54}$), both accompanied by the release of two neutrons.

Enrico Fermi and his colleagues at the University of Chicago achieved the first self-sustaining chain reaction in 1942 [10, 11]. On the site, a bronze plaque is placed that reads: "On December 2, 1942, the man got here the first self-sustaining chain reaction, and here began the controlled production of nuclear energy." The success of the experiment was announced in Washington with the following coded telephone message (the research was carried on in secret during World War II):

"The Italian navigator has just landed in the New World. The Earth was not as large as he had supposed" (meaning that the pile of uranium and graphite needed to bring about the reaction was smaller than anticipated), "so he arrived earlier than expected." Fermi called CP-1 (Chicago Pile number 1) his experimental reactor. The term nuclear reactor did not exist at the time and the experimental apparatus was called the "pile". Fermi described it as a huge mass of black and brown bricks; brown bricks were uranium and black ones the graphite blocks that were used to slow down neutrons (see the following Section).

3.7 The Slowing Down of Neutrons

Neutrons released in fission reactions are fast. As shown in the Fig. 3.6, they have energies between 0.1 and 10 MeV, with a mean value below 2 MeV. But, as shown in Fig. 3.4, fission in fissile nuclei is most readily caused by slow neutrons (energy about 0.025 eV). Then, it is evident that they must be slowed down to the thermal energy range before they have a significant probability of causing a fission and produce a chain reaction.

In the nuclear reactors, the slowing down of neutrons is obtained by letting them diffuse through a substance of low atomic weight, called *moderator*, to whose nuclei neutrons transfer their kinetic energy in a series of elastic collisions. If the target nucleus has a mass comparable to that of the neutron, the latter scatters and loses speed, which is instead acquired by the target nucleus (case (a) in Fig. 3.7). Just think of two ping-pong balls; when they collide, they can exchange much speed very readily. Instead, the exchange of speed is only partially effective if the target nucleus has a much greater mass: the neutron scatters without significantly losing energy, as it happens to a ping-pong ball that hits a bowling ball (case (b) in Fig. 3.7).

Fig. 3.6 Energy distribution of neutrons produced by the fission reaction. Obtained with data from [12]

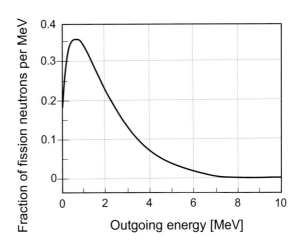

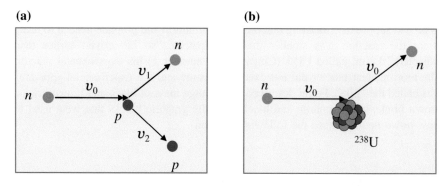

Fig. 3.7 (**a**) Elastic collision of a neutron with a hydrogen nucleus (a proton): as the mass of the proton is slightly less than that of the neutron, after the collision the two particles have almost equal speeds (and energies), (**b**) Elastic collision of a neutron with a much heavier nucleus (uranium): the neutron scatters without significantly losing energy

It can be shown (see Problem 3.10) that in an elastic collision of a neutron, of kinetic energy T_0, with a nucleus of mass number A, at rest, the fraction of the energy transferred to the nucleus (that is the ratio of the final kinetic energy of the nucleus, T_A, to the initial kinetic energy of the neutron, T_0), is given by the relation

$$\frac{T_A}{T_0} = \frac{4M_n M_A}{(M_n + M_A)^2} \cos^2 \varphi, \qquad (3.13)$$

where M_n and M_A are the masses of the neutron and the target nucleus, respectively, and φ is the angle at which the nucleus recoils.

The ratio of the final to initial neutron kinetic energies, T_n/T_0, is

$$\frac{T_n}{T_0} = 1 - \frac{4M_n M_A}{(M_n + M_A)^2} \cos^2 \varphi. \qquad (3.14)$$

With a good approximation $M_A \cong A M_n$, and Eqs. (3.13) and (3.14) can be rewritten

$$\frac{T_A}{T_0} = \frac{4A}{(1 + A)^2} \cos^2 \varphi, \qquad (3.15)$$

$$\frac{T_n}{T_0} = 1 - \frac{4A}{(1 + A)^2} \cos^2 \varphi. \qquad (3.16)$$

The maximum energy loss that a neutron can undergo occurs in a head-on collision ($\varphi = 0$). From Eq. (3.15) it follows that the energy exchange is most

Table 3.3 Energy transfer in a head-on elastic collision of neutrons with nuclei of mass number A, at rest

Target nucleus	^1H	^2H	^4He	^{12}C	^{238}U
A	1	2	4	12	238
T_A/T_0	1	0.89	0.64	0.28	0.017

efficient for $A = 1$, that is for collisions with hydrogen, for which $4A/(1 + A)^2$ is highest, and becomes very inefficient for $A \gg 1$.

Table 3.3 gives the energy transfer in a head-on collision with different target nuclei at rest; as it is seen T_A/T_0 decreases rapidly with increasing the mass M_A of the target nucleus. In particular, in a head-on, elastic collision with hydrogen at rest, the neutron stops and transfers all its kinetic energy to the proton, which moves forward with the neutron's initial speed. Instead, in the collision with a uranium nucleus, the neutron transfers only less than 2 % of its kinetic energy to the target.

Since uranium has very large mass, more than 230 times larger than that of the neutron, in a nuclear reactor the collisions inside the fuel do not slow down neutrons. Therefore, the reactor geometry is designed in such a way as to make the neutrons travel, between one fission and the other, in a medium containing hydrogen, whose nucleus (the proton) has a mass slightly lower than that of the neutron. So, on the average, at each collision with a proton the energy of the neutrons halves (see Problem 3.11).

The average number of collisions which are necessary to reduce the energy of fission neutrons from $E_{\text{fission}} \sim 2$ MeV to the thermal energy $E_{\text{th}} \sim 0.025$ eV is about 18 in water, about 25 in heavy-water (which contains deuterium with $A = 2$ instead of hydrogen) and about 115 in graphite [13]. These are the most commonly used neutron moderators in nuclear thermal reactors. As expected water is the most efficient moderator.

Problem 3.10 Elastic collision of two balls. Consider the elastic collision of a ball of mass m, moving with a speed v much lower than that of the light (kinetic energy $T_0 = \frac{1}{2}mv_0^2$), with a ball, of mass M, at rest. Calculate the kinetic energies $T_1 = \frac{1}{2}mv_1^2$ and $T_2 = \frac{1}{2}Mv_2^2$ of two balls after the collision (here v_1 and v_2 are the velocities of the two balls). Consider the system of the two balls isolated.

Solution: The motion of the two balls takes place on a plane. Let us call (see the figure)

- p_0 and p_1 the linear momentum of the ball of mass m before and after the collision,
- p_2 and T_2 the corresponding quantities of the ball of mass M,
- ϑ and φ the angles at which the balls m and M are deflected, respectively,

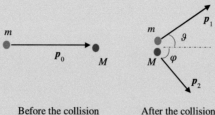

Before the collision   After the collision

Being the speed v of the ball m much lower than that of the light c, we can neglect the relativistic corrections and write

(a)
$$T_0 = \frac{p_0^2}{2m},$$
$$T_1 = \frac{p_1^2}{2m},$$
$$T_2 = \frac{p_2^2}{2M}.$$

Conservation of energy and linear momentum (by hypothesis, the system of two balls is an isolated system) gives

(b) $$T_0 = T_1 + T_2,$$

(c) $$p_1^2 = p_0^2 + p_2^2 - 2p_0 p_2 \cos\varphi.$$

The relationship (c) is shown in vector form in the following figure:

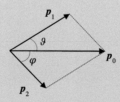

Momentum conservation in the collision

The momentum in a direction normal to the plane containing the three vectors is clearly zero.
Using the relationships (a) between kinetic energy and momentum, the last equation can be rewritten:

(d) $$mT_1 = mT_0 + MT_2 - 2\sqrt{mT_0MT_2}\cos\varphi.$$
Eliminating T_1, with the help of equation (b), we get:

(e) $$m(T_0-T_2) = mT_0 + MT_2 - 2\sqrt{mT_0MT_2}\cos\varphi ,$$
which simplifies to:

(f) $$T_2(m+M) = 2\sqrt{mMT_0T_2}\cos\varphi.$$
Squaring both sides, and dividing by $(m+M)^2T_0T_2$, we get

(g) $$\frac{T_2}{T_0} = \frac{4mM}{(m+M)^2}\cos^2\varphi,$$

which, changing T_2 with T_A, is Eq. (3.13).
From the relation (b) it is easy to obtain

(h) $$\frac{T_1}{T_0} = 1 - \frac{4mM}{(m+M)^2}\cos^2\varphi,$$

which, changing T_1 with T_n, is the Eq. (3.14).
From the above, it follows that the fraction of energy transferred to the ball of mass M is maximum for $\varphi = 0$ (head-on collision) and is equal to:

(i) $$\frac{T_2}{T_0} = \frac{4mM}{(m+M)^2}.$$

It is easily seen that this expression is highest in the case where the two balls have equal mass, $m = M$, for which $T_2 = T_0$. In the collision, the initial ball transfers all its energy to the target ball and comes to rest. In this case, the angles θ and φ are complementary ($\theta + \varphi = 90°$) (from Eq. (b) one has $v_0^2 = v_1^2 + v_2^2$) and it follows:

$$\frac{T_2}{T_0} = \cos^2\varphi = \text{sen}^2\theta,$$
$$\frac{T_1}{T_0} = \text{sen}^2\varphi = \cos^2\theta.$$

As an aside we notice that from the last figure it follows that $p_0^2 = p_1^2 + p_2^2 + 2\,p_1p_2\cos(\vartheta + \varphi)$. Using Eqs. (a), this can be written

$$2mT_0 = 2mT_1 + 2MT_2 + 2\sqrt{2mT_1 2MT_2}\cos(\vartheta + \varphi).$$

Eliminating T_0 with Eq. (b) and dividing for T_2, gives

$$m - M = 2\sqrt{mM\frac{T_1}{T_2}}\cos(\vartheta + \varphi).$$

From this equation, calling $\Theta = (\vartheta + \varphi)$ the angle between the velocities of the two balls after the collision, it follows that $\Theta < 90°$, $\Theta = 90°$, or $\Theta > 90°$ for $M < m$, $M = m$ and $M > m$, respectively.

Problem 3.11 Average energy transfer in a collision. When a ball of mass m hits a ball of mass M at rest, the latter recoils at an angle φ which is always $\leq \pi/2$. With reference to Eq. (3.16), show that after a scattering off a nucleus with mass number A, on the average, the kinetic energy of a neutron changes according to the ratio

$$\left\langle\frac{T_n}{T_0}\right\rangle = \frac{1+A^2}{(1+A)^2}.$$

Solution: The average value of the Eq. (3.16) over the interval $[0, \pi/2]$ of the possible values of the angle φ is defined as

$$\left\langle\frac{T_n}{T_0}\right\rangle = \frac{1}{\pi/2 - 0}\int_0^{\pi/2}\left[1 - \frac{4A}{(1+A)^2}\cos^2\varphi\right]d\varphi.$$

From well-known trigonometric formulas, we have that $\cos^2\varphi = \frac{1 + \cos 2\varphi}{2}$. Then, the previous integral is

$$\left\langle\frac{T_n}{T_0}\right\rangle = \frac{2}{\pi}\int_0^{\pi/2}\left[1 - \frac{4A}{(1+A)^2}\frac{1+\cos 2\varphi}{2}\right]d\varphi = \frac{2}{\pi}\left[\frac{\pi}{2} - \frac{\pi}{4}\frac{4A}{(1+A)^2}\right]$$
$$= 1 - \frac{2A}{(1+A)^2},$$

which is the given relation

$$\left\langle\frac{T_n}{T_0}\right\rangle = \frac{1+A^2}{(1+A)^2}.$$

For a scattering off a hydrogen nucleus ($A = 1$), $<T_n> = T_0/2$, i.e. on the average the neutron loses half of its energy at each collision and, therefore, few collisions are sufficient to rapidly decrease its energy. On the contrary, for a scattering off a heavy nucleus $A \gg 1$, one has $<T_n> \cong T_0$, or, in other words, the neutron has to undergo many collisions in order to significantly lose energy.

By simply applying the above formula n times, we find that the average number of collisions which are necessary to reduce the energy of fission neutrons from $E_{fission} \sim 2$ MeV to the thermal energy $E_{th} \sim 0.025$ eV is about 26 in water, about 31 in heavy-water and about 119 in graphite. The difference between these numbers and those given above in the text is due to the fact that multiple collisions create a $1/E$ neutron energy distribution, while the repeated application of the above formula assumes a flat distribution.

Problem 3.12 Elastic collision of α-particles. An α-particle with speed $v = 1.0 \times 10^7$ m/s collides elastically with the following standing particles: (a) an electron, (b) a proton, (c) the nucleus of an helium atom, (d) the nucleus of a carbon atom and (e) the nucleus of a uranium-238 atom. Calculate the maximum possible speed V of the struck particle in each case, and the percentage f of the α-particle original energy that is transferred.

Solution: Let us call $T_\alpha = \frac{1}{2}mv^2$ the kinetic energy of the α-particle before the collision, and $T_A = \frac{1}{2}MV^2$ that of the target after the collision. The maximum possible speed of the struck target is given by Eq. (3.13) for an head-on collision ($\varphi = 0$)

$$f = \frac{T_A}{T_0} = \frac{MV^2}{mv^2} = \frac{4mM}{(m+M)^2}.$$

From which we obtain

$$V^2 = v^2 \frac{m}{M} \frac{4mM}{(m+M)^2} = \frac{4m^2v^2}{(m+M)^2}.$$

Then

$$V = \frac{2mv}{m+M}.$$

Substituting the relevant values of the mass of the various target particles one obtains:

(a) collision with an electron ($M \cong 5.0 \times 10^{-4}$ u): $V_e = 2 \times 10^7$ m/s; $f_e = 0.05$ %;

(b) collision with a proton ($M \cong 1.0$ u): $V_p = 1.6 \times 10^7$ m/s; $f_p = 64$ %;

(c) collision with a ^{4}He nucleus ($M \cong 4.0$ u): $V_\alpha = 1.0 \times 10^7$ m/s; $f_\alpha = 100$ %;

(d) collision with a ^{12}C nucleus ($M = 12.0$ u): $V_C = 5.0 \times 10^6$ m/s; $f_C = 75$ %;

(e) collision with a ^{238}U nucleus ($M \cong 238.0$ u): $V_U = 3.3 \times 10^5$ m/s; $f_U = 6.5$ %.

3.8   The Thermal Nuclear Reactor

A nuclear reactor is a system where a controlled release of energy is achieved through the neutron-induced fission of certain heavy elements. By the suitable arrangement of the fuel and other reactor materials, a steady self-sustaining chain reaction is maintained, in which, on the average, just one of the neutrons released by each fission goes on to cause a further fission.

There are many different types of nuclear reactors. What is common to all of them is that they produce thermal energy that can be used as such or converted into mechanical energy and ultimately, in the vast majority of cases, into electrical energy. Most existing reactors use thermal neutrons to produce fissions and, therefore, they are called *thermal reactors*. In the following, we will mainly refer to this type of standard reactors.

There are several components common to most types of thermal reactors: the fuel, the moderator, the control rods and the coolant. Together they constitute the *core* of the reactor, where fission occurs and thermal energy is generated.

3.8.1   The Fuel

It is a component, usually in the solid state, consisting of fissile elements (e.g. ^{233}U, ^{235}U, ^{239}Pu, ^{241}Pu), where the fission reactions and most of the energy transformation of fission energy into thermal energy occurs.

Uranium is the basic fuel in most of the operating reactors. The only isotope existing in nature that can fission at any energy of the neutron (in particular at higher probability for $T_n \rightarrow 0$ MeV) is ^{235}U, but it represents only 0.720 % of natural uranium, as extracted from the ores, the rest being essentially the isotope ^{238}U (99.275 %). This low percentage is not, in general, sufficient to maintain a chain reaction. Therefore, as fuel it is customary to use uranium enriched in ^{235}U up to a percentage of between 3 and 5 % in weight (the so-called *enriched uranium*), in the form of oxide pellets UO_2 of approximately 1 cm radius, in turn contained in cylindrical metallic rods few metres long (so-called *fuel rods*). In turn, several fuel rods are put together via a suitable mechanical support structure to form a *fuel assembly* (see Fig. 4.9).

Many reactors use mixed-oxide (MOX) fuel—a mixture of uranium dioxide and plutonium dioxide. The plutonium dioxide mainly results from the commercial recycling of spent fuel. Other possible reactor fuels are thorium, which is a fertile material that produces fissile ^{233}U after neutron absorption and transmutation; uranium salts which can be used in liquid metal reactors; and other forms of uranium like uranium nitrides or uranium carbides.

3.8.2 The Moderator

It is the material used to slow down (to *moderate*) the neutrons released in fission, so that they cause more fissions. It must contain light atoms, because, as seen in the Sect. 3.7, neutrons lose energy more effectively in collisions with light nuclei, and must have a low neutron absorption cross section for fast and slow neutrons. Ordinary water (H_2O) (so-called *light-water*) has two hydrogen nuclei, which have essentially the same mass as a neutron. Therefore, as shown in Sect. 3.7, it is an ideal medium to slow them down, and in fact it is generally employed in reactors using enriched uranium. However, light-water exhibits a non-negligible neutron absorption, which reduces the number of neutrons available for fission. On the contrary, with reactors using natural uranium, the only suitable moderators are graphite and *heavy-water* (i.e. deuterium-rich water), which have low levels of unwanted neutron absorption.

The moderator is missing in fast reactors that use fast neutrons to fission and therefore do not need to slow down fission neutrons (see Sects. 3.11 and 4.4).

3.8.3 The Absorber

The neutron multiplication between two successive fission reactions can be controlled by mixing a substance to the moderator (such as boron, cadmium, hafnium, and gadolinium) which absorbs neutrons without undergoing fission. The greater the amount of absorber introduced, the smaller the multiplication of neutrons and the slower the development of the chain reaction.

The absorber is usually assembled in a few-metres long cylindrical tubes, called *control rods*. The actual amount of absorber, and thus the development of the reaction, can be varied by increasing or decreasing, with mechanical movements, the length portion of the control rods immersed in the moderator.

3.8.4 The Coolant

It is a fluid circulating through the core used to limit its temperature and, in power reactors, to transfer the thermal energy from the core to the steam driving the turbines that produce electricity. The coolant could be water, heavy-water, liquid sodium, helium gas, carbon dioxide, liquid lead or a liquid lead-bismuth eutectic mixture. In water or heavy-water reactors, the moderator functions also as coolant.

The core of the reactor is surrounded by the *reflector*, a material (usually water, graphite, or beryllium) that reflects back most of the neutrons directed towards the exterior, thus decreasing the leakage of neutrons through the walls of the core.

Usually a robust, dome-shaped steel vessel contains the reactor core and the moderator/coolant. It is designed to protect the reactor from outside intrusion and to avoid dispersion of radioactive material in case of accidents, like core melting due to insufficient cooling. The vessel is in turn embedded in a *biological shield*, typically a metre-thick concrete and steel structure designed to protect outside workers from radiation emitted by the core, especially during operation.

Figure 3.8 shows a simplified view of a reactor core, which helps to visualize the sequence of events taking place in the reactor between two successive fissions. The core of the reactor is represented as a pool of water in which both the fuel rods and the control rods are immersed. Using a single fast neutron as an arbitrary starting point (the neutron far left on the figure), we see how this fast neutron fissions a ^{235}U nucleus and produces two fragments and three fast neutrons. The fission fragments collide with the surrounding atoms within the fuel rods, slow down and stop in thousandths of a millimetre. In this process, they give up their kinetic energy as thermal energy, pretty much as, in a sudden braking, a car warms up the soil and tires. The result is a heating of the fuel rod, whose heat is then taken away by the coolant fluid.

The three fast neutrons produced in the first fission event escape from the fuel rods and, as they travel outward, they collide with the atoms of the moderator, losing energy in the process. Their speed decreases and the probability for them to

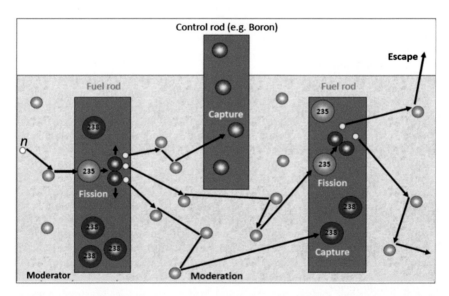

Fig. 3.8 Schematic description of the processes occurring in the core of a reactor: the fuel rods are shown as *green rectangles*, and the control rods as *grey rectangles* immersed in the moderator (*light blue*). The *spheres* with the numbers 235 and 238 represent the nuclei of isotopes ^{235}U and ^{238}U; those with *dark-blue* edge represent fission fragments; those with *purple* and *grey* edge are the nuclei of the moderator and absorber, respectively. The smaller *white circles* represent neutrons. The *arrows* indicate the path of the neutrons, which change direction, following the collisions with the nuclei of the moderator

split another fissile nucleus, within another fuel road, increases (it is the case of the middle-emitted neutron in the figure. In that fission, two further neutrons are emitted). In their path, the neutrons may also encounter the control rods; when this occurs, they are absorbed and cannot cause further fissions. They may also be captured by a nucleus ^{238}U in a fuel rod, giving rise to the formation of a radioactive nucleus heavier than uranium, with long half-life, even thousands of years (so-called *minor actinides*, see Eqs. (3.10) and (3.11) for the case of neptunium-plutonium production). Neutron capture in structural materials of the reactor, and neutron escape from the volume of the core are also possible. To minimize neutron escape from the core, as mentioned above a reflector material (not shown in the figure) surrounds the core, with the purpose of kicking neutrons backward into the core by means of nuclear elastic scattering.

The coolant, flowing through the spaces between the rods, cools them, becoming itself hotter, and thus carries away the heat that is generated by the fission process.

3.9 The Physics of a Thermal Nuclear Reactor

The sequence of processes from the production of a neutron to its final absorption on a fissile nucleus is usually called a *generation* of neutrons, and the time required for this sequence to take place is called *generation time*. Naturally, individual neutrons have different life histories. In a typical nuclear power reactor system, there are about 40,000 generations of neutrons every second [14]. In a generation and a generation time one considers a suitable average over all neutrons to describe the behaviour of the system.

A detailed study of neutron transport in a reactor is a difficult problem, which is far beyond the scope of this book. Here we shall consider the problem in a very simple approximation in order to exemplify why and how the concept of a critical mass emerges.

Figure 3.9 illustrates schematically the different processes occurring to neutrons in a thermal reactor, in one generation.

Let us consider as a starting point an arbitrary, very great number N_i of fast neutrons produced in fissions occurred in the ith generation. A small fraction of these N_i neutrons will be captured by nuclei of ^{238}U before they are slowed down and will cause their fission (as shown in Fig. 3.4, fast neutrons have a non-negligible fission cross section on ^{238}U). These neutrons will produce additional fission neutrons that will concur to sustain the chain reaction. We shall call this *fast-fission factor*, ε, where obviously $\varepsilon > 1$.

The resulting εN_i fast neutrons will diffuse in the moderator, and slow down regularly to the thermal energy. During the slowing down process, some neutrons are lost in radiative capture on ^{238}U (as shown in Fig. 3.4, this reaction has a significant cross section in the resonance region), or because they escape the reactor. We shall call P_f the fraction of neutrons that are *not* captured, and L_f the fraction of neutrons that do *not* escape from the reactor. Clearly, the fraction of

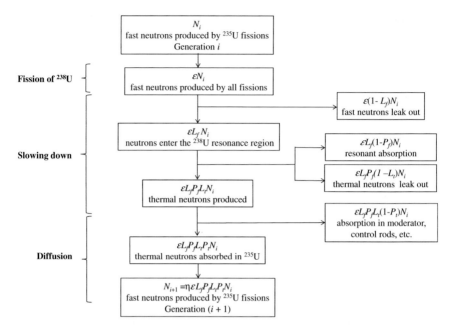

Fig. 3.9 Schematic representation of the life-cycle of neutrons in a thermal reactor

neutron absorbed is $\varepsilon(1 - P_f)N_i$, and the fraction of those that are lost is $\varepsilon(1 - L_f)N_i$. Then, the neutrons that reach the thermal energy and enter the thermalized phase (that is move around keeping, on average, the same kinetic energy) are $\varepsilon L_f P_f N_i$.

Some fraction $(1 - L_t)$ of these thermal neutrons also leak out through the walls of the reactor and another fraction P_t are absorbed in the moderator or in other structural materials. Here, we have called L_t the fraction of thermal neutrons that do *not* escape the reactor and P_t the fraction of thermal neutrons that are *not* captured by the reactor materials. This finally leaves $\varepsilon L_f P_f L_t P_t N_i$ thermal neutrons to cause fission of ^{235}U.

If η is the average number of neutrons emitted per thermal neutron absorbed (in fission and other reactions) in the fuel, at the end of the cycle there will be $N_{i+1} = \eta \varepsilon L_f P_f L_t P_t N_i$ fast neutrons available for a new cycle. The number η is called *fission factor*. It is smaller than the average number of neutrons produced per fission, v, because not all neutrons that are absorbed by the uranium produce fissions. The two quantities are related by the following relationship

$$\eta = v \frac{\sigma_f}{\sigma_f + \sigma_r}, \tag{3.17}$$

where σ_f is the fission cross section and σ_r is the absorption cross section for thermal neutrons by all processes, excluding fission (mostly the radiative capture reaction, (n, γ)).

From the above it follows that in a cycle the number of neutrons increases by a factor of

$$k_{\text{eff}} = \eta \varepsilon L_f P_f L_t P_t. \tag{3.18}$$

This parameter is called the *effective multiplication factor*. It is essentially the ratio of the number N_{i+1} of thermal neutrons in any generation to the number N_i of thermal neutrons in the previous generation.

The condition $k_{\text{eff}} = 1$ is the *criticality condition* for a finite reactor; when $k_{\text{eff}} < 1$, the reactor is subcritical, and when $k_{\text{eff}} > 1$, it is supercritical.

Equation (3.18) contains two factors, L_f and L_t, which account for the leakage of neutrons through the walls of the reactor. Clearly, they depend on the geometry and size of the reactor, including the type and size of the reflector surrounding the fuel elements. The other four factors, instead, depend only on the composition and nature of the fuel, the moderator and the other materials. It is convenient to consider all these factors separately and introduce the parameter k_∞, which is called the *infinite multiplication factor*,

$$k_\infty = \eta \varepsilon P_f P_t. \tag{3.19}$$

The Eqs. (3.18) and (3.19) are known as the "six-factor formula" and the "four-factor formula", respectively.

Clearly, $k_\infty > k_{\text{eff}}$, as $k_{\text{eff}} = L_f L_t k_\infty$, and both L_f and L_t are less than unity. Then, to attain criticality, one must compensate for the losses of neutrons by using materials with a k_∞ sufficiently larger than unity.

Of the four factors on the right side of Eq. (3.19), only η can be greater than unity. The two factors P_f and P_t are always less than unity, and $\varepsilon \cong 1$. However, also in the most favourable cases, η is not much greater than unity. It depends on the fuel, and is a function of the neutron energy. This is shown in the Fig. 3.10, which gives the average η as a function of neutron energy for the fissile nuclei ^{233}U, ^{235}U, ^{239}Pu and ^{241}Pu.

From the above it follows that, once the composition of the fuel (and then the value of η) is fixed, one must choose the moderator, its quantity and geometry, as well as the other components of the reactor, such as to make the product $P_f P_t$ as high as possible. In particular, it is necessary to avoid using materials with a high absorption cross section for thermal neutrons, which obviously reduces the factor P_t.

Once the composition of the reactor and its geometry are fixed in order to have $k_\infty > 1$, imagine to vary its dimensions, i.e. consider a series of similar finite reactors but of different sizes. Clearly, the factors L_f and L_t will depend on the size of the reactor: a large reactor will have a smaller leakage, as leakage roughly depends from the ratio surface (square of the dimension) to volume (cube of the dimension) and is therefore inversely proportional to the size. In order to avoid the L_f and L_t factors to decrease too much, a reflector surrounding the core will also have to be introduced as part of the design.

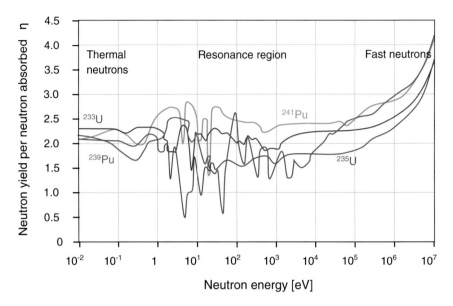

Fig. 3.10 Number of newly generated neutrons per absorbed neutron, η, as a function of the neutron energy for the different fissile nuclei.

As the L_f and L_t factors grow, tending to unity, with the increase of the size of the reactor, there will be a certain *critical size* for which $L_f L_t = 1/k_\infty$, and then, according to Eq. (3.18), $k_{eff} = 1$. The reactors with a size smaller than this critical one cannot work. It turns out that the smaller k_∞ the greater the critical size; for $k_\infty = 1$, the critical size would be infinite. This means that great care must be exercised in the choice of materials such as to make k_∞ as large as possible and consequently to make the critical size smaller and, therefore, the construction of the reactor more cost-effective.

3.10   Reactor Control and Delayed Neutron Emission

The behaviour of a nuclear reactor is governed by the distribution in space, energy and time of the neutrons in the system. In a critical reactor, the number of neutrons of a generation equals that of the previous generation and the power delivered is steady. This condition is achieved, for any level of thermal power, only for a defined ratio between the mass of the fuel, that of the moderator and the one of the absorber, that is of the immersed part of the control rods.

To decrease (increase) the output power one has just to make the reactor sub-critical (supercritical) by slightly inserting (withdrawing) the control rods in the moderator to increase (decrease) the absorption of neutrons, returning them back to the critical position once the desired power is reached. Thus, control rods are also used to maintain the reactor in a critical state over a long term.

The ability to control the chain reaction is completely due to the presence of the small proportion of delayed neutrons arising from fission. Without these, any change in the critical balance of the chain reaction would lead to a virtually instantaneous and uncontrollable rise or fall in the neutron population.

We have seen in Sect. 3.9 that, on average, in a generation the number of neutrons increases by a factor k_{eff}. Then, denoting as τ the average generation lifetime, and taking into account that one neutron is used for maintaining the chain reaction, the neutron density in the reactor grows according to the equation

$$\frac{dN}{dt} = N \frac{k_{\text{eff}} - 1}{\tau}, \tag{3.20}$$

whose solution is

$$N = N_0 e^{(k_{\text{eff}} - 1)t/\tau}, \tag{3.21}$$

where N_0 and N are the neutron density at the time zero and at the time t. The factor $\tau/(k_{\text{eff}} - 1)$ in the exponent of Eq. (3.21) is called the *reactor time constant* (or *reactor period*) and is indicated by the letter T_R.

In a properly operating reactor, k_{eff} is close to unity. Since the typical lifetime of prompt neutrons in thermal reactors is of the order of 10^{-4} s, even for values of the multiplication factor as low as $k_{\text{eff}} = 1.001$, the neutron population would increase by more than a factor of twenty thousand in one second [$e^{(1.001-1) \times 1/0.0001} = 22{,}026$]. This would make any reactor impossible to control.

Fortunately, delayed neutrons increase the effective average lifetime of neutrons to nearly 0.1 s, so that in one second, in a core with $(k_{\text{eff}} - 1) = 10^{-3}$, the neutron population would increase by only a 1 % [$e^{(1.001-1) \times 1/0.1} = 1.01$]. This is a controllable rate of change.

Most nuclear reactors are, then, operated in a condition which is subcritical if only prompt neutrons are considered, but critical if also delayed neutrons are taken into account: the prompt neutrons alone are not sufficient to sustain a chain reaction, but the delayed neutrons make up the small difference required to keep the reaction going on. This state is called a *delayed critical condition*.

This has a crucial impact on how reactors are controlled: when a small portion of control rod is slid in or out of the reactor core, the power level changes at first very rapidly due to prompt subcritical multiplication and then more gradually, following the exponential growth or decay curve of the delayed critical reaction. Furthermore, increases in reactor power can be performed at simply by withdrawing a sufficient length of control rod.

The control rods are also used to maintain the reactor in a critical state over a long term. With the progress of burnup (this is the amount of electricity generated from a given amount of fuel, see Sect. 3.12), the fissile material in the fuel rods decreases and fission products accumulate, which decreases the effective neutron multiplication factor. When continuous operation is necessary, the fuel is usually loaded so that $k_{\text{eff}} > 1$ and then adjusted to $k_{\text{eff}} = 1$ by inserting the control bars. When the amount of fissile material has decreased and fission products have accumulated with the progress of burnup, some of the control bars can be removed. This will be described with more details in Sect. 3.12.

3.10.1   Reactivity

Under normal operating conditions, a reactor is operating at or very near criticality, i.e. k_{eff} is nearly equal to 1. Since one has typically to deal with values of k_{eff} very close to unity, it may be more practical to introduce a new quantity, called *reactivity*, indicated by the symbol ρ, and defined by

$$\rho = \frac{k_{\text{eff}} - 1}{k_{\text{eff}}}. \tag{3.22}$$

It follows immediately that for a critical reactor, $\rho = 0$. In fact, since k_{eff} is very close to unity, ρ can be well approximated by

$$\rho = (k_{\text{eff}} - 1) = dk_{\text{eff}}. \tag{3.23}$$

For example a multiplication factor $k_{\text{eff}} = 1.003$ gives a $dk_{\text{eff}} = 0.003$.

Like k_{eff}, also the reactivity is a quantity that is well defined mathematically in neutron transport theory, but cannot be directly measured in practice. Being derived from k_{eff}, the reactivity depends the same way on many factors like fuel composition, core geometry, etc. As an example, let us consider the dependence of the reactivity on the core temperature. This dependence arises because the fission cross section (i.e. the probability of a neutron to produce a fission reaction) depends on the relative kinetic energy of the neutron-nucleus system. In turn, such relative kinetic energy depends on the temperature of the medium. Obviously, there is more sensitivity to temperature changes when neutrons have thermal energies, or in the epithermal regime where it is critical to be on top of a specific resonance peak or even slightly away from the peak. Instead, when the kinetic energy is much higher, say above 100 keV, it becomes less sensitive to the temperature. By looking back at Fig. 3.4, one can easily see that, at thermal energies, an increase in core temperature, corresponding to an increase in kinetic energy, will produce a decrease in fission cross section. This will in turn cause a decrease in k_{eff} or, equivalently, a decrease in reactivity. This is referred to as the nuclear composition of the fuel having a *negative temperature feedback* or a *negative temperature coefficient*.

There are of course many coefficients in reactor physics, corresponding to several changes in the reactor status. For each of them there will be a positive or negative feedback mechanisms. Therefore any changes in the reactor status can in principle produce a change in reactivity that has to be evaluated and simulated very carefully when designing a reactor, with the goal of having an overall negative feedback when power increases spontaneously (see Sect. 5.4).

3.11   Fast Reactors

An alternative to the thermal reactor strategy is to use primarily fast neutrons to sustain the fission chain reaction. This is done in a fast reactor, where neutrons are used without moderation, which for uranium is less efficient than using slow neutrons.

Various fuels and combinations of fuels can provide the required self-sustaining reaction with fast neutrons. For instance, highly enriched uranium (over 20 % ^{235}U) can be used because at this concentration of U-235, fissions are sufficient to sustain the chain-reaction despite the smaller probability of fission with fast neutrons. A more efficient fuel is fissile plutonium or a mixture of uranium and plutonium. In the latter case, the U-238 will produce additional plutonium.

In a uranium reactor part of the neutrons are absorbed by U-238 and ultimately yield Pu-239 and Pu-241. The reactor thus "converts" some U-238 into Pu-239 and Pu-241. These nuclei are fissile nuclei and undergo fission in the same way as U-235 releasing thermal energy. Equations (3.10) and (3.11) show the reactions that convert ^{238}U into ^{239}Pu.

The ratio of desired, fissile nuclei produced (through transformation of fertile nuclei) to the fuel nuclei burned up (both by fission and other reactions) is called the *conversion ratio* and is usually indicated by the letter C,

$$C = \frac{\text{Fissile nuclei produced}}{\text{Fissile nuclei destroyed}}. \tag{3.24}$$

Let us examine in more details the operation of a fast reactor. Suppose we have initially w natural uranium atoms in a reactor. Without using the conversion, one can burn at the maximum all the U-235 content, namely an amount $0.007w$ (we have assumed a 0.7 % of ^{235}U in natural uranium).

With a reactor with a conversion ratio C, after having burned all the uranium-235, one has produced a quantity $0.007wC$ of plutonium. Then, one can operate such a reactor to burn the produced plutonium and have the residue U-238 as fertile material to be converted. Together with burning the above amount of plutonium, a quantity $0.007wC^2$ of plutonium will be produced by conversion, which can again be burned, and so on. After n stages, the total amount of fissionable material from which energy has been produced will be

$$0.007w\left(1 + C + C^2 + \ldots + C^n\right) = 0.007w\,\frac{1 - C^{n+1}}{1 - C}. \tag{3.25}$$

If $C < 1$, for n large this becomes

$$0.007w\frac{1}{1-C},\tag{3.26}$$

and this will be the maximum amount of fissile material that can be burned. It is a factor of $1/(1-C)$ times larger than the amount that can be burned without conversion.

From Eq. (3.26) it is easy to conclude that it suffices a small conversion factor to largely increase the energy output. For example, for $C = 0.8$, which is a valid figure for a heavy-water moderated reactor, the obtainable energy increases by a factor of five. If a reactor could be operated so that all the fissionable material, both original and produced in operation, were consumed, it would be possible, assuming $C = 0.8$, to obtain five times greater power production than it would be provided by the original concentration of ^{235}U only. However, such fuel recycling cannot be achieved by simply putting the used fuel back again into a reactor. This is because the fission products, present in the fuel, have typically a high neutron absorption cross section and, therefore, would hinder the operation of the reactor by decreasing the available number of neutrons. Hence, the above sequence of fuel cycles can only be implemented if the used fuel is reprocessed to remove the fission products. This is called "reprocessing of the spent fuel". A nuclear reactor which generates fissile material, but less than it uses is called a *converter reactor*.

An interesting case is when $C > 1$, that is the reactor produces by conversion more fissile nuclei than it burns. In this case, the amount of fissile material produced in subsequent stages grows and there is no limit to the process except the exhaustion of all available fertile material. In this case the conversion process is called *breeding* and has a great economic importance, because it allows to burn completely the natural uranium, thus multiplying, in principle, by a factor of $1/0.007 = 140$ the energy that can be produced. A reactor which generates more fissile material than it uses is called *breeder reactor*.

The conversion factor of Eq. (3.24) can be approximated as

$$C = \eta - P - k,\tag{3.27}$$

where η is the average number of neutrons released per neutron absorbed in the fuel (see Fig. 3.10), P is the number of neutrons that are lost because they escape the reactor or are absorbed in the various other materials of the reactor, and k is the number of neutrons absorbed by the fissile material which keep the chain reaction. In a critical condition $k = 1$, and the relationship (3.27) becomes

$$C = \eta - P - 1.\tag{3.28}$$

From this equation it follows that to achieve a good conversion ratio ($C > 1$) it is, obviously, necessary to reduce the losses P, making the reactor large, and avoiding the use of neutron-absorbing materials. It is also necessary that

Table 3.4 Spectrum averaged η [14]

Neutron spectrum	Pu-239	U-235	U-233
Light-water reactor (LWR)	2.04	2.06	2.26
Oxide fuelled fast reactor	2.45	2.1	2.31

$\eta > (2 + P)$, where the term P has to made as small as possible. From the Fig. 3.10 it is seen that Pu-239 offers the highest conversion ratio at high energies due to the higher value of η. Uranium-233 is significantly better than uranium-235 and it could feature C higher than unity even in thermal reactors. In Table 3.4 are given the values of η averaged over the neutron energy spectrum for these three fissile isotopes and for thermal and fast reactors. From this table it is clear that, even though η is slightly greater than 2 for Pu-239 or U-235 at thermal energies, thermal breeders are practically difficult to achieve with these fissile fuels due to the neutron losses by parasitic absorption and leakage. In addition, the table shows that it is possible to construct thermal breeders with U-233 as fuel and that Pu-239 is a better fuel for fast reactors than U-235.

A fast reactor in which the net change of plutonium content is negative is called a *fast burner reactor*, while one in which the plutonium content is increasing is termed a *fast breeder reactor* (FBR).

The conventional fast reactors built so far are generally fast breeder reactors implying a net increase in Pu-239 from breeding, due to a conversion ratio above 1.0. They feature a *fertile blanket* of *depleted uranium* (i.e. uranium with less than the natural 0.72 % content of U-235) around the core, and this is where much of the Pu-239 is produced. The blanket can then be reprocessed (as are the fuel elements in the core) and the plutonium recovered for use in the core, or for further fast nuclear reactors.

3.11.1 Doubling Time

The rate of production of fissile material compared to its initial inventory is quantified by a term called *doubling time*, usually denoted as *DT*. It is the operating time required to produce excess fissile material equal to that loaded initially in a reactor system. Doubling time is one of the figures of merit used to compare breeder reactor designs, different fast reactor fuels, and fuel cycle systems involving many breeder reactors. Clearly, for faster growth of fissile material, a short *DT* is desired.

The doubling time *DT* can be defined as,

$$DT = \frac{\text{Initial Fissile Mass}}{\text{Net Fissile Mass Production Rate}}. \tag{3.29}$$

There are different types of doubling times depending if one considers only the reactor, the system (i.e. the reactor and its support facilities) or also the external fuel cycle (i.e. the overall cycle of fuel and energy production). The simplest *DT* is the so-called *reactor doubling time*, denoted as *RDT*, where one assumes that the net bred fissile material is removed continuously and stored until it is equal to the initial mass. The *RDT* is the time required for a breeder reactor to produce enough fissile fuel, in excess of its own initial fissile inventory, to fuel an identical reactor. Hence, it is the time necessary to double the initial load of fissile fuel. It does not account for the fuel cycle external to the reactor. It is defined as

$$RDT = \frac{M_0}{\dot{M}_g},\qquad(3.30)$$

where M_0 is the initial fissile inventory (usually measured in kg) and $\dot{M}_g$ (in kg/a) is the excess fissile inventory produced per year. From the basic core physics properties, *RDT* can be estimated using the following equation [15]:

$$RDT \cong \frac{2.7M_0}{G(1+\alpha)Pf},\qquad(3.31)$$

where $G = (C - 1)$ is the breeding gain, P is the reactor thermal power in megawatts (MW), α is the capture to fission ratio and f is the fraction of reactor operating time at nominal power (*load factor*). To obtain a low *RDT*, a high power density (P/M_0) and a large breeding gain are desired. It is worth noticing that *RDT* is the minimum time for doubling as it has not considered the fissile fuel losses due to fabrication, reprocessing and nuclear decay.

3.12   Fuel Burnup

The total energy released in the fission of a given amount of nuclear fuel is called fuel *burnup*. It is measured in megawatt-days[6] (abbreviated as MWd). The fission energy release per unit mass of the fuel is termed *specific burnup* of the fuel and is usually expressed in megawatt-days per kilogram of the heavy metal originally contained in the fuel (meaning actinides, like thorium, uranium, plutonium, etc.), abbreviated as MWd/kgHM, where HM stands for heavy metals. Often used is the practical unit megawatt-days per tonne (MWd/tHM).

   During irradiation in a nuclear power plant, the composition and physical characteristics of the fuel change continuously from its initial physical properties, and this has important consequences on the fuel cycle. At the end of its life in the reactor, the fuel needs to be characterized as accurately as possible for reasons related to both safety and economy. The main investigations carried out are oriented

---

[6]A Megawatt-day is a quantity of energy corresponding to a power of 1 MW ($=10^6$ J/s) extended over one day ($=86,400$ s). Therefore 1 MWd $= 8.64 \times 10^{10}$ J. Another similar unit is the Gigawatt-day (GWd), corresponding to a power of 1 GW over one day, equal to 8.64×10^{13} J.

towards verifying the fuel cladding integrity and determining the fissile content and the fuel burnup, the latter being an indicator of the fuel cycle efficiency.

Regarding burnup, one usually distinguishes between *low burnup fuel* (<30 GWd/t) and *high burnup fuel* (>60 GWd/t), meaning fuels that have been in the reactor for a short and long time, respectively. The distinction between the two categories is quite arbitrary and depends on the advances in fuel technology. In fact, the maximum burnup value, evaluated when the used fuel is discharged from the reactor, is always moving upwards.

The burnup is the main parameter that one can vary to modify the energy production of a reactor without redesigning it. However, increasing the burnup implies an extension of the in-core time of the fuel, and this has complex practical consequences which need careful analysis before a decision is taken.

The nuclear reactions occurring in the reactor while the fuel is being burned produce two main categories of new radioactive elements:

(a) Transmuted elements that have captured one or more neutrons without disintegrating and are, therefore, heavier than fissile nuclei (transuranic elements). These elements are part of the group of actinides (Np, Pu, Am, Cm and others).
(b) Fission products, which are much lighter than the original nuclei. Some of these elements are in the gaseous state. They and their decay products form a significant component of nuclear waste.

Both these categories of materials, accumulating in the fuel rods with the progress of burnup, tend to prevent the correct developing of the chain reaction by decreasing the effective neutron multiplication factor. Therefore, periodically and according to different core management strategies (see also Sect. 4.7 and Ref. [16]), typically in 3 years the whole core is downloaded and replaced by fresh fuel. The irradiated fuel extracted from the reactor is then managed and treated in different ways, depending on the adopted fuel cycle (see Chap. 4).

The fuel extracted from the reactor after use constitutes the radioactive waste. It is the so-called *spent nuclear fuel* (SNF).

3.12.1 Transmutation

As the fuel is irradiated in the reactor, the quantities of fissile materials in the fuel change significantly. This change is referred to as transmutation, i.e. conversion of one radionuclide into another through one or more nuclear reactions. The analysis of these changes is important for the optimal use of each fuel assembly in the nuclear power plant.

The calculation of the actual conversion of nuclides into other nuclides is complicated since: (*a*) the neutron flux is space, energy and time dependent; (*b*) the neutron capture cross-section is energy dependent; and (*c*) the various atoms are distributed heterogeneously in the core.

In thermal Light-Water Reactors (LWRs), the most important fissile materials are U-235 and several isotopes of plutonium. There is no U-235 production, and U-235 depletion occurs because of neutron absorption (including both fission and capture). Since most of the fissions are U-235 fissions (some are Pu-239 and ^{238}U fissions), a reactor operating with a constant power output will have close to a constant loss rate of U-235. This implies that, as the number of U-235 atoms in the core decreases, the neutron flux will have to increase to maintain a constant power output.

Figure 3.11 shows the consumption of the fissile U-235 and fertile U-238 while new fissile Pu-239 and Pu-241 (as well as some fission products and other acti-nides) are produced through radiative capture in fertile U-238 and Pu-240. The figure relates to a typical Pressurised Water Reactor (PWR) plant core. Different reactors give somewhat different curves. In this graph the production or loss of several isotopes in kilogram per tonne of uranium (kg/tU) is plotted as a function of burnup, measured in gigawatt-days per tonne of uranium (GWd/tU). This graph shows that the U-235 losses are not quite linear with burnup. The loss rate decreases at higher burnup because the fissile plutonium isotopes (the isotopes with odd mass number) become more important in producing power.

3.12.2   Plutonium Isotope Production

The build-up of the various plutonium isotopes is more complicated than that of the U-235 since there is both production and loss of the isotope. Part of the newly formed plutonium (one-third to one-half) undergoes fission in the reactor, and thus contributes to energy production. The remainder eventually leaves the reactor with the rest of the nuclear waste.

Plutonium-239 is formed by a neutron capture reaction in U-238 and two sub-sequent beta radiations emissions (see Eqs. 3.10 and 3.11) and is lost through neutron absorption. Since the half-life of Pu-239 is very long (24,110 a), its radioactive decay is negligible compared to the burnup and further transmutation.

Figure 3.11 shows how Pu-239 builds up rapidly during the initial fuel irradi-ation and then levels off to almost an equilibrium value (about 5 kg/tU) at high burnups. Its production is relatively constant over the irradiation period as can be seen from the steady slope of the U-238 curve in Fig. 3.11. The loss rate of Pu-239 is proportional to the amount of this isotope in the core. Thus initially the pro-duction rate far exceeds the loss rate, but as burnup proceeds the loss rate increases until it just balances the production rate.

Other isotopes of plutonium are formed by successive neutron capture reactions. Capture in Pu-239 yields Pu-240. Capture in Pu-240 yields Pu-241, etc. The actual amounts of these various isotopes in individual spent fuel assemblies will vary depending on the irradiation history of each fuel assembly.

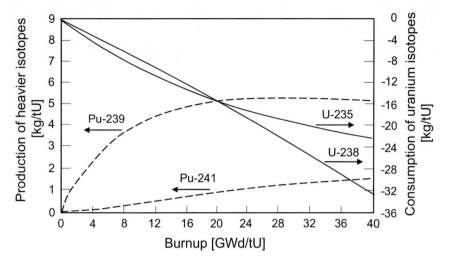

Fig. 3.11 Consumption of ^{235}U and ^{238}U, and production of new fissile ^{239}Pu and ^{241}Pu as a function of burnup in a PWR plant. Pu-239 and Pu-241 production are measured by the left-hand vertical scale, while U-235 and U-238 consumption are measured by the right-hand vertical scale

3.12.3 Fission Fragments

As discussed in Sect. 3.4, the fission reaction produces a broad spectrum of isotopes whose mass numbers are concentrated in two intervals around $A = 95$, the lighter fragment, and around $A = 138$, the heavier fragment (see Fig. 3.3). Most of these fission fragments are radioactive and undergo a series of decays to form stable end products.

Of particular importance for reactor control and operation are fission fragments and their decay products which, having a large absorption cross-section, are likely to absorb neutrons. The build-up of such fission products is one of the factors in determining when reactor fuel elements must be removed from the reactor and replaced. Two such fission products that play an important role in reactor plant operations are Xe-135 and Sm-149.

Let us examine in detail the case of Xe-135 as an example. The genetic relations in the $A = 135$ chain are shown in the Fig. 3.12. The isotope ^{135}Xe is formed either directly by fission with a 0.6 % yield, or through the decay of other direct fission products, ^{135}Te and ^{135}I, which are rather common fission products (fission yields of 3.5 and 2.5 %, respectively).

Iodine-135 has a rather small probability for absorbing a neutron, so it plays a negligible role in reaction rate control. It decays, with a half-life of about 6.6 h, into xenon-135 which in turn decays, with half-life 9.1 h, into ^{135}Cs. The xenon-135 has

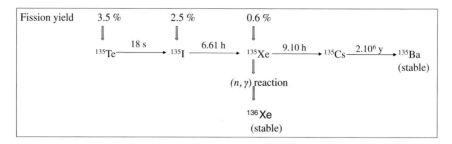

Fig. 3.12 Production of ^{135}Xe in a fission reaction

a very large cross-section for neutron absorption, about 3 million barns under reactor conditions. This compares to 400-600 barns for the uranium fission event.

During normal operation of a nuclear reactor at a constant neutron flux, the large radiative cross section of ^{135}Xe continuously converts it into ^{136}Xe, thereby keeping the amount of ^{135}Xe small and in equilibrium with that of ^{135}I. However, after a strong reduction in neutron flux, ^{135}Xe formation by decay of ^{135}I is more rapid than its conversion into ^{136}Xe. Then the amount of ^{135}Xe increases and consequently the reactivity drops. If the reactor has been shut down for a while, it may be impossible to restart it until the amount of ^{135}Xe has been reduced again by decay (this effect is called *xenon poisoning*). However, if the reactor can be restarted the ^{135}Xe will be converted into stable ^{136}Xe and the reactivity will rapidly increase until a new concentration equilibrium is reached. This is called *xenon transient* and can temporarily make the reactor difficult to control. A major contribution to the sequence of events leading to the Chernobyl nuclear disaster (see Sect. 5.9.5) was the failure to anticipate the effect of xenon poisoning on the rate of the nuclear fission reaction in the nuclear reactor.

Safety concerns focus on the possibility of release of fission fragments into the environment. In this area, the isotopes of greatest concern are those which have high activity and which can cause damage if inhaled or ingested. Among the most important of these isotopes are Sr-90, Cs-137, and several isotopes of iodine (this will be discussed in detail in Chap. 6).

Problem 3.13 Rate of consumption of nuclear fuel. A nuclear electric power plant produces heat by fission of ^{235}U. Its efficiency to convert heat into electricity is 35 %, and its capacity for electricity production is $P_{el} = 1000$ MW(e).

(a) Calculate the rate of consumption of ^{235}U, in nuclei per second, knowing that 8.19×10^{10} J are released from the fission of 1 g of ^{235}U (see Problem 3.3).

(b) If there are initially 2000 kg of ^{235}U in the reactor, estimate the approximate time when the fuel will have to be replaced.

Solution:

(a) At 35 % efficiency, the thermal capacity of the power plant is

$$P_{th} = \frac{P_{el}}{0.35} = \frac{1000}{0.35} = 2857\,\text{MW}.$$

Assuming that the fission of 1 g of U-235 releases approximately 8.19×10^{10} J of kinetic energy, the consumption of uranium per second is:

$$R = \frac{2.857 \times 10^9}{8.19 \times 10^{10}} \approx 0.035\,\text{g/s of }^{235}\text{U}.$$

(b) The time T for complete consumption of the initial 2000 kg of uranium is, then

$$T = \frac{2,000,000}{0.035} = 5.7 \times 10^7\,\text{s} = 1.8\,\text{a}.$$

Problem 3.14 Power production, rating and fuel consumption. A thermal reactor, which contain $M = 150$ tonnes of natural uranium, runs with a neutron flux $\Phi = 10^{13}$ cm^{-2} s^{-1}. Knowing that fission and capture cross section on ^{235}U are $\sigma_f = 579$ b and $\sigma_c = 101$ b, respectively, determine the developed power, the rating (power per mass unit) and the fuel consumption. Assume that the energy released per fission is $E = 200$ MeV ($= 3.2 \times 10^{-11}$ J).
Solution: The percentage of ^{235}U nuclei in natural uranium (mass number about 238) is 0.72 %, then the number $N(^{235}\text{U})$ of ^{235}U nuclei in the mass M of fuel is

$$N\left(^{235}\text{U}\right) = \frac{1.5 \times 10^5 \times 0.0072 \times 6.022 \times 10^{23}}{0.238} = 2.73 \times 10^{27}\,\text{nuclei}.$$

Therefore, the reactor develops the power

$$P = N\left(^{235}\text{U}\right)\Phi\,\sigma_f E = 2.73 \times 10^{27} \times 10^{13} \times 579 \times 10^{-24} \times 3.2 \times 10^{-11}$$
$$= 505.8\,\text{MW},$$

with the rating

$$\frac{P}{M} = \frac{5.06 \times 10^8\,\text{W}}{150\,\text{t}} = 3.37\,\text{MW/t}.$$

The consumption of ^{235}U nuclei is due to both fission and capture processes. Then, observing that the total absorption cross section of ^{235}U is

$$\sigma_a = \sigma_f + \sigma_c = 579 + 101 = 680\,\text{b},$$

the number of ^{235}U burned per second is

$$N\left(^{235}\text{U}\right)\sigma_a \Phi = 2.73 \times 10^{27} \times 680 \times 10^{-24} \times 10^{13} = 1.86 \times 10^{19}\,\text{s}^{-1}.$$

In one year ($\sim 3.15 \times 10^7$ s), about 5.9×10^{26} nuclei are burned. This is about 20 % of the initial load of ^{235}U.

Problem 3.15 Fuel breeding. Consider a fast breeder reactor operating with the conversion ratio $C = 1.3$. Determine the amount of pure plutonium fuel to be burnt, in order to accumulate an additional 1500 kg of fissile material. *Solution*: By definition of the conversion factor of Eq. (3.24) it follows that

$$C = \frac{\text{number of fissile nuclei produced}}{\text{number of fissile nuclei destroyed}} \cong \frac{\text{mass of fissile nuclei produced}}{\text{mass of fissile nuclei destroyed}}.$$

Then, calling M the mass of fissile nuclei consumed, one has

$$C = \frac{M + 1500}{M} = 1.3.$$

From this relation it follows that $M = 5000$ kg. Hence, 5000 kg of plutonium must be burnt to produce an additional 1500 kg of fissile material.

Problems

3.1   A target consists of 200 identical nails hammered into the surface of a slab of softwood 2 m^2 area. The nail heads act as hard object that can scatter small bullets (assume their size is negligible) fired at the target; those hitting the wood pass through undeflected. Calculate the effective area of the nail head if, out of a total of 10^5 bullets sprayed uniformly over the slab, only 500 are deflected.
[*Ans.*: $5.0 \cdot 10^{-5}$ m^2]

3.2   In a scattering experiment, a calcium foil containing 1.3×10^{19} nuclei per cm^2 is bombarded with a 10 nA beam of α particles, and 9.5×10^4 protons per second are emitted. Determine the total cross section for the reaction ^{48}Ca $(\alpha, p)^{51}$Sc.
[*Ans.*: 1.169×10^{-25} cm^2]

3.3 Calculate the energy that a neutron loses in a head-on collision with: (a) a hydrogen nucleus, (b) a nitrogen-14 nucleus, and (c) a lead-206 nucleus.
[*Ans.*: (a) 100 %; (b) 25 %; (c) 1.9 %]

3.4 A beam of protons of $I = 5 \times 10^7$ s^{-1} intensity bombards a target of thallium-210 of 0.1 g/cm^2 thickness. Calculate the number of neutrons produced in the reaction: $p + {}^{210}\text{Tl} \rightarrow {}^{210}\text{Pb} + n$, knowing that the cross section of the process is $\sigma = 50 \times 10^{-6}$ b.
[*Ans.*: 0.7 s^{-1}]

3.5 A beam of neutrons, 0.29 eV kinetic energy and 10^5 s^{-1} intensity, bombards a foil of ^{235}U 0.1 kg/m^2 thickness. Calculate the number of fission reactions occurring per second in the foil, knowing that the fission cross section at the given neutron energy is 200 b.
[*Ans.*: 512 fissions per second]

3.6 A beam of neutrons of kinetic energy 0.1 eV bombards a cube of natural uranium 1 cm side. The neutron flux is 10^{12} s^{-1} cm^{-2}, and the density of uranium is 18.9 g/cm^3. Estimate the rate of heat generation in the cube due to the fission of ^{235}U (natural abundance 0.72 %), knowing that the fission cross section of ^{235}U at that energy is 250 b.
[*Ans.*: 2.75 W]

3.7 Determine the nature of the nucleus ^{A}X produced in the following fission reaction

$$n + {}^{235}\text{U} \rightarrow {}^{138}\text{Xe}_{54} + {}^A\text{X} + 2n.$$

[*Ans.*: ^{96}Sr$_{38}$]

3.8 Estimate the energy released in the fission reaction of Eq. (3.7): $n + {}^{235}\text{U}_{92} \rightarrow {}^{137}\text{I}_{53} + {}^{96}\text{Y}_{39} + 3n$, knowing that the atomic masses of the nuclei ^{235}U, ^{137}I$_{53}$ and ^{96}Y$_{39}$ are 235.043923 u, 136.917871 u and 95.915891 u, respectively.
[*Ans.*: 179.6 MeV]

3.9 Estimate the energy released in the fission reaction $n + {}^{235}\text{U}_{92} \rightarrow {}^{141}\text{Ba}_{56} + {}^{92}\text{Kr}_{36} + 3n$, knowing that the atomic masses of the nuclei ^{235}U, ^{141}Ba$_{56}$ and ^{92}Kr$_{36}$ are 235.043923 u, 140.914411 u and 91.926156 u, respectively. Calculate the total fission energy of 1 kg of ^{235}U$_{92}$ assuming all fissions proceed via this reaction.
[*Ans.*: 173.3 MeV, about 7.10×10^{13} J]

3.10 In a nuclear reactor 3×10^{19} fissions per second occur. Estimate the thermal power output of the reactor, assuming that on the average the energy release per fission is 200 MeV.
[*Ans.*: 960 MW]

3.11 A nuclear reactor runs on uranium-235 and develops 4.6 GW of thermal power. Assuming that on an average 200 MeV of energy is released per fission of a single ^{235}U nucleus, estimate the nuclear fuel burn rate of the reactor.
[*Ans.*: 5.6×10^{-5} kg/s = 0.056 g/s]

3.12 Assuming that on an average 200 MeV of energy is released per fission of a single ^{235}U nucleus, calculate the amount of fuel burned per month in a nuclear-powered submarine traveling at an average power of 30 MW.
[*Ans.*: 948 g]

3.13 Burnup of 1 kg of carbon would release about 3×10^7 J of thermal energy. Estimate how many kg of carbon must be burned to produce the same energy as for the fission of 1 kg of ^{235}U (assume an average energy release per fission of 200 MeV).
[*Ans.*: 2.73×10^6 kg]

3.14 A nuclear power plant that develops 1.0 GW(e) of electric output has a 33 % efficiency in converting the thermal energy into electric energy. Calculate the mass of ^{235}U burned per day, assuming that on an average 200 MeV of energy is released per fission of a single ^{235}U nucleus.
[*Ans.*: 3.16 kg/d]

3.15 A nuclear reactor consumes 20.4 kg of U-235 in 1000 h of operation. Determine the power developed by the reactor assuming that on an average the energy released per fission is 200 MeV.
[*Ans.*: 465 MW]

3.16 A nuclear power plant developing 800 MW(e) of electric output has a 33 % efficiency in converting the thermal energy into electric energy. The load factor of the plant is 75 %. Calculate the mass of ^{235}U burned per year, assuming that on an average 200 MeV of energy is released per fission of a single ^{235}U nucleus.
[*Ans.*: 700 kg/a]

3.17 The cooling water of a pressurised water reactor enters the core at 216 °C temperature and exits at 287 °C. The water is under pressure and does not transform into vapour. The core develops 5×10^9 W thermal power. Determine the mass of water that passes though the core of the reactor, knowing that, in the specific temperature range, the specific heat of water is 4420 J/(kg °C).
[*Ans.*: $\sim 1.6 \times 10^4$ kg/s]

3.18 A sodium ^{23}Na target is bombarded by a constant current of 10^{-5} A of 14 MeV deuterons. The deuterons undergo the stripping reaction $d + {}^{23}\text{Na} \rightarrow p + {}^{24}\text{Na}$, forming the sodium isotope ^{24}Na, which is radioactive and undergoes β^- decay (half-life $\tau_{1/2} = 14.8$ h) into ^{24}Mg. The yield of the reaction (i.e. rate of production of activity of the ^{24}Na isotope) is 4.07×10^9 Bq per hour. Calculate (a) the maximum activity of the isotope ^{24}Na that can be produced under these bombardment conditions; (b) the activity of ^{24}Na produced in 8h of continuous bombardment; (c) the activity of ^{24}Na eight-hour after the conclusion of the bombardment in (b).
[*Ans.*: (a) = 8.69×10^{10} Bq, (b) 2.72×10^{10} Bq, (c) 1.87×10^{10} Bq]

References

1. J. Chadwick, *Possible existence of a neutron*, Nature, vol. 129, p. 312, and the existence of a neutron, in Proceedings of the Royal Society A, vol. 136, p. 692 (1932)
2. E. Segrè, Spontaneous fission. Phys. Rev. **86**, 21 (1952)
3. N.E. Holden, D.C. Hoffman, *2000 IUPAC, Pure and Applied Chemistry*, vol. 72, No 8 (2000), pp. 1525–1562
4. J.K. Shultis, R.E. Faw, *Fundamentals of Nuclear Science and Engineering* (CRC Press, 2008), pp 141 (table 6.2). ISBN 1-4200-5135-0
5. https://www.oecd-nea.org/janis/
6. O. Hahn, F. Strassmann, On the detection and characteristics of the alkaline earth metals formed by irradiation of uranium with neutrons. Die Naturwissenschaften **27**, 11 (1939)
7. L. Meitner, R.O. Frisch, Disintegration of uranium by neutrons: a new type of nuclear reaction. Nature **143**, 239 (1939)
8. http://wwwndc.jaea.go.jp/cgi-bin/FPYfig?xpar=a&zlog=unset&typ=g2&part=a
9. Java-based Nuclear Data Information System, http://www.oecd-ne.org/janis/
10. E. Fermi, *Fermi's Own Story*. The first reactor (United States Atomic Energy Commission, Division of Technical Information, Oak Ridge, 1942), pp. 22–26. Report OCLC 22115
11. C. Allardice, E.R. Trapnell, *The First Pile*, United States Atomic Energy Commission Report TID 292 (1946)
12. http://www.oecd-nea.org/janisweb/tree/N/BROND-2.2/DE/U/U235/MT18
13. J.L. Basdevant, J. Rich, M. Spiro, Fundamentals in Nuclear Physics (Springer, Berlin, 2005), p. 301. ISBN 0-387-01672-4
14. J.A. Angelo, *Nuclear Technology* (Greenwood Publishing Group, 2004), p. 516. ISBN 1-57356-336-6
15. A.E. Waltar, A.B. Reynolds, *Fast breeder reactors* (Pergamon Press, New York, 1981). ISBN 978-0-08-025983-3
16. S. Glasstone, A. Sesonske, *Nuclear Reactor Engineering* (Springer), ISBN 978-1-4613-5866-4

Part II
Energy from Nuclear Fission

Chapter 4
Nuclear Reactors

This Chapter is about the existing and future nuclear power plants. After a classification of nuclear reactors according to the purpose, the type of nuclear fuel, moderator and operation mode, we briefly describe the features common to each generic type of reactors, providing data on the number of reactors in operation and under construction worldwide as well as reviewing the peculiarities of the main conventional designs. The chapter also deals with the nuclear fuel cycle processes including the analysis of the world reserves and demand of nuclear fuel.

4.1 Classification of Nuclear Reactors

Nuclear reactors are classified according either to the purpose or to the type of the nuclear fuel, the operation mode—which specifies whether a moderator is used and the type of heat transfer agent employed—and the arrangement of fuel, moderator and coolant.

In terms of the purpose, nuclear reactors are subdivided into *power reactors* or *research reactors*. Power reactors are mainly used in nuclear power plants for electricity production. Their lesser uses are propulsion of large ships and submarines, heating, desalination of seawater or brackish, conversion of fertile material into fissile material. The reactors of this category are always of considerable thermal power, up to a few gigawatts. Research reactors are operated at universities and research centres in many countries, including some where no nuclear power reactors are operated. These reactors generate neutrons for multiple purposes, including producing radioisotopes for medical diagnosis and therapy, testing materials and conducting basic research. The reactors of this category have generally modest power, from a few kilowatts to a few megawatts.

In terms of operation mode, the reactors are classified as *fast*, *intermediate*, and *thermal reactors*, depending on the speed of the neutrons causing fission. In fast reactors, neutrons are used almost as they are produced by fission, so most of the

© Springer International Publishing Switzerland 2016   147
E. De Sanctis et al., *Energy from Nuclear Fission*, Undergraduate Lecture
Notes in Physics, DOI 10.1007/978-3-319-30651-3_4

fission reactions take place with fast neutrons (>100 keV). In thermal reactors, neutrons are slowed down to thermal energies, and most of the fissions are with low-energy (thermal) neutrons. In intermediate reactors, neutrons are slowed down to energies in the middle (range of 0.5 eV to thousands of electron volts).

With respect to the arrangement of the nuclear fuel, moderator (if present) and coolant (if different from the moderator) in the core, nuclear reactors are classified as *homogeneous*, in which these two substances are finely divided and uniformly mixed together, and as *heterogeneous*, in which the two substances are in separate elements as blocks.

4.2 Nuclear Power Plants

Since the early 1950s, nuclear power has developed to provide up to about 15 % of the world's electricity. Nuclear power has seen significant improvements in reactor life span, capacity and safety records; but not without challenges. Economics, nuclear weapons proliferation concerns, nuclear waste and safety are four main issues that influence the use and growth of nuclear power around the world.

The principles that allow for the use of nuclear power to produce electricity are the same for most types of reactor. The energy released from continuous fission of the nuclei of the fuel is harnessed as thermal energy in a fluid (water, gas, liquid, metal or molten salt), and is used to produce steam. The steam is used to drive the turbines which produce electricity, as in a fossil-fuel-fired thermoelectric plant.

Figure 4.1 schematically shows the basic components of a nuclear power plant based on a pressurised water type of reactor; this is similar to any other thermo-electric plant except that the source of the heat producing the steam that drives the steam turbines is the nuclear reactor core, rather than the fossil fuel furnace.

Like thermal power plants, a nuclear plant reactor cannot convert 100 % of its thermal energy into electricity. Then, the power output of a nuclear reactor is quoted in two ways:

- Thermal megawatt or MW(th), which is the nominal thermal power output in megawatt of the nuclear plant. It depends on the design of the specific nuclear reactor itself and relates to the quantity and quality of the steam it can produce;
- Megawatt electric or MW(e), which is the electric output capability of a power plant in megawatt. It is equal to the thermal overall power multiplied by the efficiency of the plant.

Clearly, the closer a power plant's MW(th) and MW(e) ratings are, the more efficient it is. Typically the electric output of a nuclear power plant is one third of the thermal power. Therefore the typical efficiency of the power generation in a nuclear power plant is 30–33 %, even though some nuclear reactors (e.g. fast or gas-cooled reactors) can reach efficiencies as high as 40 %. For comparison, the efficiency of modern coal-, oil- or gas-fired power plants reaches up to 40 %.

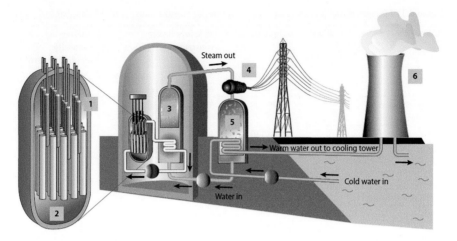

Fig. 4.1 Basic components of a thermal nuclear power reactor (pressurised water reactor): *1* Reactor: fuel rods (*light blue*) heats up pressurised water. Control rods (*grey*) absorb neutrons to control or halt the fission process. *2* Coolant and moderator: fuel and control rods are surrounded by water (primary circuit) that serves as coolant and moderator. *3* Steam generator: water heated by the nuclear reactor transfers thermal energy through thousands of pipes to a secondary circuit of water to create high-pressure steam. *4* Turbo-generator set: steam drives the turbine, which spins the generator to produce electricity just like in a fossil-fuel plant. *5* Condenser: removes heat to convert steam back to water, which is pumped back to the steam generator. *6* Cooling tower: removes heat from the cooling water that circulates through the condenser, before returning it to the source at near-ambient temperature. Cooling towers are needed by some plants to dump the excess thermal energy that cannot be converted into mechanical energy due to the laws of thermodynamics. It is worth noticing that they emit only clean water vapour [1]

As of 31 October 2015, there were 439 nuclear power reactors in operation (i.e. connected to the electricity grid), in 30 countries, with an installed electric net capacity (i.e. the available nominal power) of about 380 GW(e) [2, 3]. Also at that date, there were 69 nuclear power plants under construction with an installed capacity of more than 65 GW(e).

In Table 4.1 are listed the countries where there are nuclear power reactors in operation and/or under construction, as of October 2015. The Table also gives the relevant power of the plants and the nuclear electricity supplied in 2014. The entire Japanese nuclear reactor fleet (which amounts to 43 units) is still considered in operation or operational. In reality, because of the Fukushima Daiichi nuclear accident on March 11, 2011 (see Sect. 5.9), no nuclear power has been generated in Japan since September 2013 and many of the operational units will likely never generate any power again. At the moment of writing this book, two units (the Sendai-1 and Sendai-2 reactors) have been restarted and resumed commercial operation in September and November 2015, respectively [4]. More units are expected to follow in the upcoming months.

As mentioned above, the nominal generating capacity as of October 2015 was 380.065 GW(e). However, for the actual electricity production one must take into account the reactors in temporary shutdown and the so-called *load factor*, which

Table 4.1 Nuclear power plants in operation and under construction, as of 31 October 2015 [3]

Country	Reactors in operation		Reactors under construction		Nuclear electricity supplied in 2014	
	No of units	Total MW (e)	No of units	Total MW (e)	TWh	% of tota
United States of America	99	98,708	5	5633	798.6	19.5
France	58	63,130	1	1630	418.0	76.9
Japan	43	40,290	2	2650	0.0	0.0
Russia	34	24,654	9	7371	169.1	18.6
China	29	25,025	23	22,738	123.8	2.4
Republic of Korea	24	21,677	4	5420	149.2	30.4
India	21	5308	6	3907	33.2	3.5
Canada	19	13,500			98.6	16.8
United Kingdom	16	9373			57.9	17.2
Ukraine	15	13,107	2	1900	83.1	49.4
Sweden	10	9651			62.3	41.5
Germany	8	10,799			91.8	15.8
Belgium	7	5921			32.1	47.5
Spain	7	7121			54.9	20.4
Czech Republic	6	3904			28.6	35.8
Taiwan	6	5032	2	2600	48.8	18.9
Switzerland	5	3333			26.5	37.9
Finland	4	2752	1	1600	22.6	34.7
Hungary	4	1889			14.8	53.6
Slovakia	4	1814	2	880	14.4	56.8
Argentina	3	1627	1	25	5.3	4.1
Pakistan	3	690	2	630	4.6	4.3
Brazil	2	1884	1	1245	14.5	2.9
Bulgaria	2	1926			15.0	31.8
Mexico	2	1330			9.3	5.6
Romania	2	1300			10.8	18.5
South Africa	2	1860			14.8	6.2
Armenia	1	375			2.3	30.7
Iran	1	915			3.7	1.5
Netherlands	1	482			3.9	4.0
Slovenia	1	688			6.1	37.3
Belarus			2	2218		
United Arab Emirates			4	5380		
Total	**439**	**380,065**	**69**	**65,827**	**2410**	

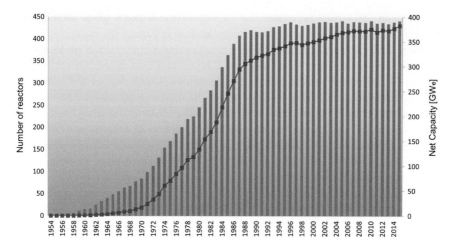

Fig. 4.2 Worldwide net nuclear generating capacity (*red curve* and right-hand vertical scale) and number of operating reactors (histogram and left-hand vertical scale) from 1954 to 2015 [5]

represents the effective operation time of a working plant. Indeed, if we look at the total production in 2014, nuclear plants worldwide generated 2410 TWh of electricity. If one multiplied the total nominal generating capacity (excluding Japan, that is $380,065 - 40,290 = 339,775$ MW(e)) by 8760 h in a year one would obtain 2976 TWh. Therefore, the actual produced electricity is about 81 % of this net nominal capacity, which means that the typical load factor is quite high.[1] In that year, 13 countries relied on nuclear energy to supply at least one-quarter of their total electricity. The first five nuclear generating countries—by rank, the USA, France, Russia, Republic of Korea and China—generated over two thirds of all nuclear electricity in the world and two countries alone—the USA and France—accounted for half the global nuclear electricity.

Electricity produced as of the end of 2014 was 11 % of the world total [5] and is still of the same order today. As illustrated in Fig. 4.2, since the 1950s, when the first commercial reactors started to appear, until 1985 the number of operating reactors as well as the nuclear net generating capacity has been increasing about linearly with time. Then, since the early 1990s, the total number of operating nuclear plants has remained nearly constant, with the retirement of older units and an equal number of new plants coming on-line. Nuclear capacity worldwide continued to increase, albeit with a reduced growth rate after 1990, mainly due to the increased size of new plants and power upgrading of existing plants.

The use of nuclear energy remains limited to a small number of countries. Close to half of the world's nuclear energy countries are located in the European Union

---

[1]Clearly, 81 % is a worldwide average load factor. For instance, in 2015 the 99 US nuclear power reactors achieved a record 91.9 % load factor, surpassing 91.8 % capacity reached in 2007.

(EU) and in 2014 they accounted for 34.5 % of the world's gross nuclear production, with half the EU generation being located in France [5].

Over 70 % (50 units) of the sixty-nine reactors under construction are located in Asia and Eastern Europe, of which half (23) are in China alone. Over fifty-five percent (38 units) are located in just three countries, China, Russia and India.

In addition to commercial nuclear power plants, there are about 240 research reactors, operating in 56 countries, with more under construction. They are used for several purposes, including research and the production of medical and industrial isotopes, education and training. Moreover, some 180 nuclear reactors propel more than 140 ships, mostly submarines, and over 13,000 reactor-year of experience has been gained with marine reactors [6].

Although a variety of plant configurations have been tested on an experimental basis since the 1950s, essentially all of the electrical power generated by nuclear power plants is produced by primarily six different plant types. These include:

- the Pressurised Water Reactor (PWR),
- the Boiling Water Reactor (BWR),
- the Pressurised Heavy-Water Reactor (PHWR),
- the Light-Water cooled, Graphite moderated Reactor (LWGR),
- the Gas-Cooled Reactor (GCR),
- the Liquid-Metal Fast Breeder-Reactor (LMFBR).

A majority of these reactors are light-water cooled (277 PWR + 15 LWGR and 80 BWR units). The other reactors in the world are: pressurised heavy-water reactors (49 PHWR units, mostly CANDU type[2]), gas-cooled reactors (15 GCR, in the UK only) and 2 liquid-metal cooled fast-breeder reactors.

Historically, the choice of a particular plant design by a manufacturer or by an electric utility has been influenced by many factors, such as the available materials (fuel and otherwise), the available technological capabilities, and the previous experience of the designers or users.

In the USA, for example, the first commercial power plants were PWR and BWR plants largely due to their similarity to the naval propulsion reactors developed shortly after World War II. The availability of a uranium enrichment capability and manufacturing capability has also influenced the design choices. In Canada, the PHWR technology was developed, utilizing natural uranium and smaller pressure vessels. Fast breeder reactors are being designed in an effort to extend the energy potential of the uranium and thorium resources of the world and to reduce the burden of the high-level radioactive wastes. The fuel for these reactors normally comes from the reprocessed fuel of light-water reactors or other reactors.

The general features of each of these plant types are given in the Table 4.2.

---

[2]The CANDU (short for CANadian Deuterium Uranium) reactor is a Canadian-invented, pressurised heavy water reactor used for generating electric power. The acronym refers to its deuterium-oxide (heavy water) moderator and its use of (originally natural) uranium fuel.

Table 4.2 The general features of different nuclear power plant types in operation

Plant type/Feature	PWR Pressurised light-water moderated and cooled reactor	BWR Boiling light-water cooled and moderated reactor	PHWR Pressurised heavy-water moderated and cooled reactor	LWGR Light-water cooled, graphite moderated reactor	GCR Gas-cooled, graphite moderated reactor	(LM)FBR Fast Breeder reactor
Fuel material	LEU[a] MOX[b]	LEU[a] MOX	Natural uranium LEU[a]	LEU[a] with thorium	Natural uranium	Plutonium mixed with natural or depleted uranium
Fissile material production	Converter	Converter	Converter	Converter	Converter	Breeder
Moderator material	Light-water	Light-water	Heavy-water	Graphite	Graphite	None
Neutron energy	Thermal	Thermal	Thermal	Thermal	Thermal	Fast
Reactor coolant	Light-water	Light-water	Heavy-water	Light-water	Carbon dioxide	Liquid metal
Number in Operation	278	80	49	15	15	2
Main countries	USA, France, Japan, Russia, China	USA, Japan, Sweden	Canada	Russia	UK	Russia

[a]LEU stays for Low Enrichment Uranium

[b]MOX stays for Mixed OXide fuel, usually consisting of plutonium blended with natural uranium, reprocessed uranium, or depleted uranium

Table 4.3 Acronyms of different nuclear reactors

ABWR	Advanced Boiling Water Reactor
ACR	Advanced CANDU Reactor
AGR	Advanced Gas-cooled Reactor
AP600	Advanced Passive-600 Pressurised Water Reactor
AP1000	Advanced Passive-1000 Pressurised Water Reactor
APWR	Advanced Pressurised water Reactor
BWR	Boiling Water Reactor
CANDU	CANadian Deuterium Uranium reactor. Canadian type of PHWR
EPR	Evolutionary Pressurised Reactor
ESBWR	Economic Simplified Boiling Water Reactor
FBR	Fast Breeder Reactor
GCR	Gas-Cooled Reactor
GFR	Gas Fast Reactor
HTGR	High Temperature Gas Reactor
HTR	High Temperature Reactor
HWR	Heavy-Water Reactor
IRIS	International Reactor Innovative and Secure
LFR	Lead-cooled Fast Reactor
LMFBR	Liquid-Metal Fast-breeder Reactor
LWGR	Light-Water Graphite Reactor
LWR	Light-Water Reactor
MSBR	Molten Salt Breeder Reactor
PBMR	Pebble Bed Modular Reactor
PBR	Pebble Bed Reactor
PHWR	Pressurised Heavy-Water Reactor
PWR	Pressurised Water Reactor
RBMK	Reaktor Bolszoj Moszcznosti Kanalnyj. Russian type of LWGR
SCWR	Super-Critical Water Reactor
SFR	Sodium Fast Reactor
VHTR	Very High-Temperature Reactor
VVER	Vodo-Vodjanoi Energetičesky Reaktor. Russian type of PWR

The Table 4.3 gives the abbreviations in common use for the different nuclear reactors.

4.3   Comparison of Various Electricity Generating Technologies

It may be interesting to compare various ways to generate 1000 MW of electricity on a continuous basis. This corresponds to the power of one large power plant. Clearly in the case of intermittent sources, like wind and photovoltaic, the 1000 MW(e) on a continuous basis are actually to be understood as average power.

Table 4.4 Comparison of fuel demand per year, greenhouse gas (GHG) emissions (measured in g/kWh of CO_2 equivalent emission), and land use for generating 1000 MW of electricity on a continuous basis using different energy sources. Values are approximations; in cases requiring conversion from thermal energy to electricity a net efficiency of 1/3 has been assumed. For the nuclear fuel, low-grade uranium-ore (0.1 % U) is considered

Power plant	Fuel needed Ref. [7]	GHG emission (gCO_2equiv/kWh) Ref. [8]	Land use (km^2) Refs. [7, 9]
Coal fired plant	3×10^9 kg of coal per year	1001	3
Oil-fired plant	2×10^9 litres oil per year	840	3
Gas-fired plant	3×10^9 m^3 gas per year	469	
Nuclear power plant	1×10^8 kg of uranium ore per year, or 1×10^5 kg of natural uranium per year, 1.5×10^4 kg of enriched fuel per year, or 0.7×10^3 kg of uranium-235 per year	16	1.5
Wind turbines off-shore	1300 turbines with 90 m rotor diameter	12	50–150
Solar cells (15 % efficiency)	60 km^2 of solar cell area	46	20–50

   Table 4.4 shows this comparison considering the amount of fuel burned per year, the life-cycle emissions of greenhouse gas (GHG), and the land use, the latter corresponding to the total surfaces that must be devoted to human activities within the energy chain.
   In the case of fossil-fueled technologies, fuel combustion during operation of the facility emits the vast majority of greenhouse gas. Instead, in the case of nuclear and renewable energy technologies, the bare energy production mechanism does not contribute to the GHG emissions, while the majority of them occur during ancillary activities (e.g. in the construction of the plants). Therefore, for a correct comparison one must use the so-called *life-cycle emissions*, that include emissions associated with the manufacturing, construction, installation and decommissioning[3] (if applicable) of the plants, and with the fuel production when relevant. In other words, for nuclear plants the life-cycle GHG emissions include emissions associated with all activities in the whole plant lifetime, i.e. emissions associated with building of the plant, mining and processing the fuel, routine operation of the plant, disposal of used fuel and other waste by-products, and decommissioning. The GHG emission data shown in Table 4.4 are the median values of aggregated results from a review of the relevant literature performed by the Intergovernmental Panel on Climate Change[4] in 2011 [8].

---

[3]Decommissioning is the process whereby a nuclear power plant or other nuclear installation is dismantled to the point that it no longer requires measures for radiation protection.

[4]The Intergovernmental Panel on Climate Change (IPCC) is the international body for the assessment of climate change established by the United Nations Environment Programme (UNEP) and the World Meteorological Organization (WMO) in 1988 to provide the world with a clear

The Table shows that nuclear energy requires a relatively low quantity of fuel, because of its high energy density. Moreover, greenhouse gas emissions of nuclear power plants are among the lowest of any electricity generation method and on a life-cycle basis are comparable to wind, photovoltaic and other renewable sources not shown in the Table (GHG emissions of hydroelectric and biomass are 4 and 18 gCO_2equiv/kWh, respectively). Life-cycle GHG emissions of natural gas, oil and coal are respectively 29, 52 and 62 times greater than nuclear.

Numerous independent studies have assessed life-cycle emissions of nuclear and renewable energy technologies. Data found in different analyses vary somewhat, reflecting differences in assumptions and in some input parameter values. However, all studies indicate that nuclear energy's life-cycle emissions are comparable to renewable forms of electricity generation, such as wind and hydropower, and are far less than those of fossil-fired power plants.

The high energy density of nuclear fuel is also reflected in the land use required for the infrastructure of a power station. For a 1000 MW(e) nuclear power station, roughly 150 hectares are needed, a size that doubles for a coal or oil-fired plant of the same power, due to the fuel storage on-site. Renewable sources are more demanding: for the same installed power, some square kilometres are needed for hydropower (a large dam), a few tens of square km for solar panels, from a few tens to more than a hundred for a wind farm, and thousands for a biomass plantation [9].

When we look at the quantity of fuel transported to operate a plant, while one single truck loading the new fuel is required to annually operate a nuclear reactor, one oil tanker a week is needed for an oil-fired power station, or one trainload per day for a coal-fired power station (see Problem 6.4), or the umbilical dependence from a pipeline for a gas-fired power station.

4.4   Nuclear Reactor Technologies and Types

As seen in the Sect. 3.8, a thermal reactor comprises fuel elements, moderator, reflector, shielding, neutron absorber, coolant, and structural material.

In a typical nuclear reactor, the fuel is not in one piece, but in the form of several hundred vertical rods, each about 4 m long, and grouped in bundles called *fuel assemblies*. Several hundred fuel assemblies containing thousands of small pellets of fuel, often in the form of ceramic oxide fuel, make up the core of a reactor. For a reactor with an output of 1000 MW(e), the core would contain about 75 tonnes of uranium.

(Footnote 4 continued)

scientific view on the current state of knowledge in climate change and its potential environmental and socio-economic impacts.

Another system of rods that contain a neutron absorbing material (control rods) can move up and down in between the fuel rods. When totally inserted in the core, the control rods absorb so many neutrons, that the reactor is shut down. To start the reactor, the operator gradually moves the control rods up. In an emergency situation they are dropped down automatically.

In a new reactor with new fuel, a neutron source is needed to start the neutron chain reaction. Usually this is beryllium mixed with polonium, radium or another alpha-emitter. Alpha particles from the decay cause a knock-off of neutrons from the beryllium by turning it into carbon-12. Restarting a reactor with some used fuel may not require this, as there may be enough neutrons from heavier nuclides produced by neutron irradiation in the core and decaying by spontaneous fission to achieve criticality when control rods are withdrawn.

The reactor core sits inside a steel pressure vessel: in the case of a light-water reactor (LWR) the pressure is between 7 (for the BWR type) and 15.5 MPa (for the PWR type) so that water remains liquid even at the operating temperature of over 320 °C. Steam is formed either above the reactor core (BWR type) or in a separate component, called steam generator (PWR type).

The reflector is placed around the core to reflect leaking neutrons back to the core. For the reflector, the same material as that of the moderator is generally used.

The shielding is placed outside the reflector so that residual radiation from the core is absorbed, and the area around the pressure vessel is shielded against neutrons and γ-rays.

The basic operation of any nuclear reactor is as follows. The kinetic energy of the fission products generated by nuclear fission and interacting with the medium (fuel) is converted into thermal energy released in the coolant (usually water) of the primary circuit, which flows through the reactor core. In the most common nuclear reactor (PWR) the primary coolant enters a steam generator and exchanges heat with water flowing in a secondary circuit which, at the given temperature and pressure, becomes steam. The steam thus produced drives the turbines that generate electricity. The steam is then condensed and the coolant recycled.

The control rods are used at the start-up and shut-down of the reactor and also to control the power output. With the fuel burnup, the fissile material in the fuel decreases and fission products, which are also neutron absorbers, accumulate. These phenomena decrease the effective neutron multiplication factor. To guarantee continuous operation of the reactor, the fresh fuel is usually loaded in the core in a configuration such that the effective multiplication factor is larger than unity ($k_{eff} > 1$); the latter is then adjusted to $k_{eff} = 1$ by inserting the control rods. When the amount of fissile material has decreased and fission products have accumulated with fuel burnup, some of the control rods are withdrawn in order to keep $k_{eff} = 1$. Thus, the control rods are not only used for start-up and shut-down, but also to maintain the reactor in a critical state over a long term.

4.4.1 Light-Water Reactors (LWR)

Light-Water Reactors (usually called with the acronym LWR) are the most widely used nuclear reactors (they account for about 85 % of the current installed capacity). These reactors use low-enriched uranium or a uranium-plutonium mixture from reprocessing as fuel, and natural water (typically called light-water, as opposed to heavy-water) as the coolant. It has also the function of moderator and reflector.

LWR include Boiling Water reactors (BWR), in which the water boils in the core as it receives thermal energy from the reactor, and pressurized-water reactors (PWR), in which water is kept liquid during operation due to the very high pressure in the primary circuit (about 15.5 MPa).

4.4.2 Pressurised Water Reactors (PWR)

Pressurised Water Reactors (PWR) are the most common type of nuclear reactors accounting for about 67 % of current installed nuclear generation capacity worldwide. The design of PWR originated as a submarine power plant. The principal features of this reactor are shown in the Fig. 4.1.

In a PWR the water that passes through the reactor core to act as moderator and coolant does not flow to the turbine, but is contained in a pressurised primary loop. The vessel that encloses the reactor core is under high pressure (about 15.5 MPa, or about 150 atmosphere). This prevents the water (which is at a temperature of over 320 °C) from boiling inside the vessel; hence the reactor operates with water always in the liquid state. The high-pressure, very hot water which leaves the reactor vessel is sent to a steam generator where it exchanges thermal energy with water circulating in a separate secondary loop. The primary water leaving the steam generator is then pumped back into the core to repeat this thermal energy transfer process. Water in the secondary loop is converted into steam in a steam generator and the generated steam is sent to the turbine. The spent steam, after delivering its energy to the turbine, is condensed, cooled further and pumped back into the steam generator. The cooling of water condensed in the turbine (say, from 100 °C to about 30 °C) is carried out in the cooling tower.

The obvious advantage of the PWR design is that a leak of radioactive nuclides in the core would not transfer any radioactive contaminants to the turbine and the condenser. Indeed, only the water in the primary loop can become radioactive as it flows through the core. To enhance the safety of the system, the primary loop is confined within the reactor containment building, usually a reinforced concrete structure. This way, a leak of radioactive water from the primary loop will also remain confined within the containment building. For the leak to propagate outside the containment building, a leak should occur in both the primary loop and the building itself, which is highly unlikely.

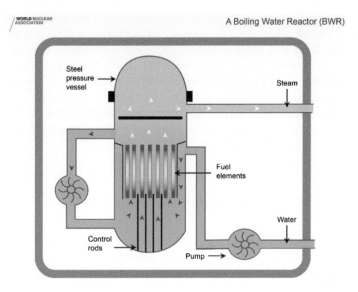

Fig. 4.3 Schematic representation of a Boiling Water nuclear Reactor (BWR). In this reactor, the same water loop serves as moderator, coolant for the core, and steam source for the turbine [10]

4.4.3   Boiling Water Reactors (BWR)

Boiling Water Reactors (BWR) are the second most common nuclear reactor type accounting for about 18 % of installed nuclear generating capacity. The Fig. 4.3 shows the essential features of a BWR.

The design has many similarities to the PWR, except that there is only a single circuit in which the water is at lower pressure (7 MPa, or about 70 atmospheres) so that it boils in the core at about 290 °C. The BWR uses only two separate water systems as it has no separate steam generator system. The reactor is designed to operate with 12–15 % of the water in the top part of the core as steam. As the steam has a lower density than liquid water, it is less effective in moderating neutrons, which decreases the overall fission efficiency of the core. This steam and water mixture passes through two stages of moisture separation. Water droplets are then removed and steam is allowed to enter the steam line and drive the turbine. Once the turbines have turned, the remaining steam is cooled in the condenser coolant system. This is a closed water system. Thermal energy from the steam is absorbed by the cool water through heat transfer. The water within the two systems does not mix. Once through the condenser system, the water is recycled back into the reactor to begin the process again.

In the BWR design, the primary loop goes outside of the containment building. Therefore, a leak in that part of the loop may lead to a spread of radioactive water that would not be as strongly confined as within the containment building. This means a greater concern about possible contaminations of the environment around the reactor. On the other hand, the presence of the steam in the upper part of the reactor vessel leads to a negative temperature coefficient: indeed, the steam is less effective in moderating the neutrons, being its density lower than that of liquid water. Therefore, since in case of overheating of the core more steam will be produced, as a consequence the neutrons will be less moderated and so less effective in causing further fissions. This means that the presence of the steam leads to a negative feedback with respect to overheating, which is a safety feature in this design.

4.4.4   Pressurised Heavy-Water Reactors (PHWR)

An alternative thermal reactor design that uses natural rather than enriched uranium is the Heavy-Water Reactor (HWR). Natural uranium contains only 0.72 % of the fissile isotope instead of the 3–4 % in the case of enriched uranium used in other reactors. Therefore, in order to maintain a comparable effective multiplication factor, a better neutron economy is required. Heavy-water serves this purpose, as the deuterium nuclei contained in it have a much lower neutron capture cross section than normal hydrogen, so that less neutrons are lost in the moderation process with respect to light-water.

The principal representative of this class of reactors is the Canadian-built CANDU reactor. More recently, the heavy-water reactor design has been developed

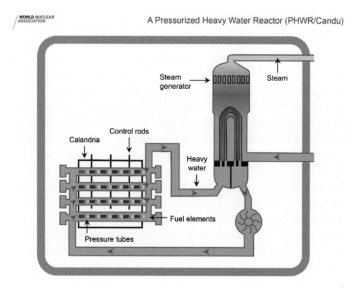

Fig. 4.4 Schematic representation of a Heavy-Water nuclear Reactor (PHWR/CANDU) [10]

in India, which has selected PHWRs for Stage I of its nuclear programme, based on economic and technical viability.

A schematic representation of the CANDU reactor is shown in Fig. 4.4. The use of natural uranium oxide as fuel avoids the cost of the enrichment process, but, as mentioned before, it requires a more efficient moderator. On the other hand, the economic benefits from avoiding the fuel enrichment process are significantly counterbalanced by the cost of heavy-water production.

Since heavy-water contains deuterium which is twice as heavy as hydrogen, it has less moderating power (see Sect. 3.7). As a consequence, the geometry of a BWR or PWR cannot be used, as much more heavy-water would be needed. Therefore, the pressure vessel design cannot be adopted in this case, which is the reason why the CANDU reactor is based on pressure tubes immersed in a heavy-water tank. This tank, called *calandria*, hosts the pressure tubes (several hundreds of them) arranged horizontally. The fuel is loaded into the tubes, where the heavy-water flows at high pressure, 10 MPa (or about 100 atmospheres), at a temperature up to 290 °C. Control rods instead penetrate the tank vertically. Besides the control rods, there is a secondary shutdown system that can intervene by adding gadolinium (a strong neutron absorber) to the moderator. The high pressure prevents the water from boiling. The temperature of the water leaving the core is not very high: in the steam generator, the steam produced is only at 260 °C at a pressure of 4.7 MPa (about 47 atmospheres). This is the reason of the low efficiency, around 28 %, of this type of reactor in generating electricity, the lowest among commercial plants.

Among the advantages of the PHWR is the fact that fuel replacement can be performed without the need of a reactor shutdown. A single pressure tube can be isolated from the cooling loop, allowing for refuelling while the reactor is still operating. Moreover, the absence of a large pressure vessel means lower construction costs for the reactor system. Reactor engineers have also come up with a different design of a Heavy-Water Reactor, the so-called Advanced Candu Reactor (ACR), where the fuel is based on slightly enriched uranium and light-water is used to cool the core.

4.4.5   Light-Water Graphite-moderated Reactors (LWGR)

The Light-Water Graphite-moderated Reactor (LWGR) is a class of reactors whose most known representative is the RBMK, abbreviation for the Russian Reaktor Bolšoj Moščnosti Kanalny, which means High Power Channel-type Reactor. This is the reactor type involved in the Chernobyl accident in 1986 (see Sect. 5.9).

The LWGR is a boiling water reactor, which uses light-water for cooling and graphite as moderator (a combination not found in any other power reactor in the world). It is very different from most other power reactor designs as it is derived from a system principally conceived for plutonium production and was intended and used in the former Soviet Union for both plutonium and power production.

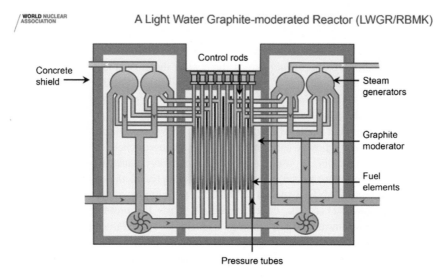

Fig. 4.5 Schematic representation of a Light-Water Graphite-moderated boiling water Reactor (LWGR/RBMK) [10]

As shown in Fig. 4.5, the design is based on pressure tubes and a single circuit, with steam generated directly in the reactor and separated in steam drums. The fuel elements are tubes sitting inside the pressure tubes, which are arranged within vertical graphite blocks whose total mass is very high. At the outlet of the steam drums the average temperature and pressure are 280 °C and 6.38 MPa (about 64 atmospheres), with a net efficiency in the conversion to electricity around 31 %.

The advantage of RBMK reactors is that they can work with a low fuel enrichment level (typically uranium dioxide with enrichment level 2.6–2.8 % in U-235) and offer the possibility to replace fuel tubes during reactor's operation (up to 5 replacements per day). However, the graphite reaches very high temperatures and if the cooling water is lost (so-called *Loss of Coolant Accident*, abbreviated *LOCA*) the neutron multiplication coefficient increases. Both these factors make RBMK reactors unsafe, as it was dramatically shown in the Chernobyl accident (see Sect. 5.9), where in addition the control rod design turned out to be inadequate. On the other hand, the chain of malfunctions that led to the disaster was apparently triggered by a specific human intervention against safety principles.

A number of significant design changes were made after the Chernobyl accident to address these problems, so that there are still more than ten of these reactors in commercial operation worldwide.

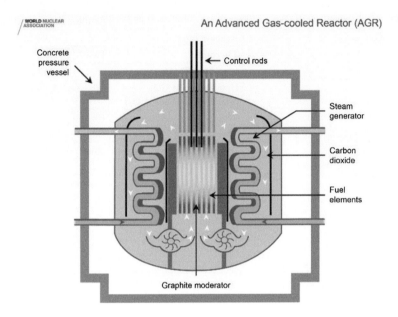

Fig. 4.6 Schematic representation of an Advanced Gas-cooled nuclear Reactor (AGR) [10]

4.4.6 Gas-Cooled Reactors (GCR)

Gas-Cooled Reactors (GCR) are nuclear reactors that use graphite as a neutron moderator and carbon dioxide gas as the primary coolant. The fuel is uranium oxide pellets, enriched to 2.5–3.5 %, in stainless steel tubes. The carbon dioxide circulates through the core, reaching 650 °C, and then is sent to the steam generator tubes outside of the core, but still inside the concrete and steel pressure vessel. Besides control rods that penetrate the moderator, there is a secondary shutdown system involving nitrogen injection into the coolant.

Figure 4.6 shows a schematic representation of an Advanced Gas-cooled Reactor (AGR). It is the second generation of British gas-cooled reactors that has been designed to have a high thermal efficiency (ratio of generated electricity to generated heat) of about 41 %, which is better than modern pressurised water reactors, which have a typical thermal efficiency of 34 %. This is due to the higher coolant outlet temperature of about 640 °C that can only be easily obtained with gas cooling, compared to about 320 °C for PWRs. However the reactor core has to be larger than light-water reactors for the same power output, and the fuel burnup at discharge is lower, so that the fuel is used less efficiently, counteracting the thermal efficiency advantage.

Table 4.5 Comparison of characteristics of typical thermal and fast reactors

Feature	Reactor type	
	Thermal	Fast
Average neutron energy	Low (0.0253 eV)	High (100–200 keV)
Fuel	Uranium-oxide UO_2 Mixed-oxide (PuO_2-UO_2)	Mixed-oxide (PuO_2-UO_2)
Fuel concentration (%)	Low (0.7–5 U-235)	High (15–20 Pu-239)
Fertile conversion factor C	Low	High
Core volume	Large	Small
Power density [kW/litre]	10	400
Coolant	Light or Heavy-water	Liquid metal
Thermal efficiency (%)	28–34	40
Fuel burnup [GWd/t]	7–40	>100
High level waste (see Chap. 6)	Produced	Partially incinerated
Neutron flux [cm^{-2} s^{-1}]	10^{14}	5×10^{15}
Maximum neutron fluence [cm^{-2}]	10^{22}	2×10^{23}

4.4.7   Fast Neutron Reactors (FNR)

All thermal reactors described above have two main disadvantages. The first is the low efficiency in fuel usage, only 1–3 % of the overall uranium element (including both U-235 and U-238), and the second concerns (except for GCR) the unsatisfactory steam parameters in the turbine circuit caused by using water as coolant in the primary circuit. Moreover, the presence of the moderator medium leads to large cores with rather low power density.

Fast neutron reactors (FNR) present a design different from thermal ones: they have no moderator and fast neutrons produced in fission reactions are not slowed down to lower energy intentionally. Accordingly, the components and materials used in the two reactor types also differ. The main difference between the characteristics of a thermal and a fast reactor are summarized in the Table 4.5.

In both thermal and fast reactors, the neutrons produced by fission have an energy distribution of the kind shown in Fig. 3.6, with an average energy of 2 MeV. As illustrated in Chap. 3, to slow down neutrons to low energies efficiently, and utilize the higher neutron fission cross-sections, the fuel in a thermal reactor is embedded in a moderator. In fast reactors, the moderator is absent in order to keep the neutron energies in the fast range to enable breeding (see Sect. 3.11). However, the neutron energy spectrum in a fast reactor will be softer than the fission spectrum due to the unavoidable moderation of neutrons mainly by elastic and inelastic scattering in the coolant and in the structural materials. Another peculiar feature of fast reactors is that, since fast neutrons have a much lower fission cross-section than thermal ones, the fuel enrichment in fissile isotopes must be higher.

The absence of moderators and the high fuel enrichment due to low cross-sections in fast reactors translates into smaller core volume and hence higher power density. Consequently, fast neutron reactors need coolants with higher heat transfer properties. In many designs, FNR are cooled using liquid metals (sodium, lead or lead-bismuth eutectic mixture) rather than water, and so are referred to as liquid-metal fast-breeder reactors (LMFBR).

In the normal operation of any reactor, besides the fission products produced by splitting uranium nuclei, neutron capture can create long-lived plutonium (from ^{238}U) and minor actinides (heavier transuranic elements like neptunium, americium, curium etc.). On one hand, plutonium production is what a breeder reactor is supposed to do, specifically creating additional fissile fuel. On the other hand, the long-lived actinides also represent radioactive waste, i.e. radiotoxic substances whose radioactivity decreases significantly only over very long periods of time (centuries or millennia) (see Chap. 6). However, since all minor actinides can be fissioned by fast neutrons, the harder neutron spectrum in fast reactors has the advantage that it also enables the incineration of such unwanted elements and hence reduces the burden of radioactive waste with respect to a thermal reactor (see further below in this Subsection).

The main components of a typical (sodium-cooled) fast reactor are:

- a suitable nuclear fuel which releases energy by nuclear fission process;
- a coolant which transfers the thermal energy to an intermediate heat exchanger;
- a secondary coolant which transfers the thermal energy from the heat exchanger to a steam generator;
- the steam generator which heats water to form steam that rotates the turbine of an electric generator to produce electricity.

The high power density (up to several hundred MW/m^3 with highly enriched fuel) and the choice of a proper coolant allow to obtain steam at temperature of 487 °C under a pressure of 17.7 MPa (about 177 atmospheres). This leads to an efficiency of the FBR reactor at the level of 40 %. Most of the proposed FBRs use sodium as coolant, a fluid with very low neutron moderating properties, as requested by a fast reactor, and excellent thermal characteristics. However, sodium activates under irradiation, which makes the fluid within the primary circuit particularly radioactive. In air, sodium can burn (in liquid form its ignition temperature is about 125 °C), while at contact with water it reacts violently and, since hydrogen is formed in the reaction, it presents an explosion hazard [11]. The combination of radioactivity and chemical reactivity imposes the adoption of a triple fluid circuit reactor. Another issue is the sodium melting point of 98 °C: to avoid its solidification sodium has to be heated up even when the reactor is under shutdown conditions, e.g. during maintenance.

Fast neutron reactors mostly use plutonium as their basic fuel, or sometimes highly-enriched uranium (20–30 % fissile nuclei) to start them off. The fuel in a

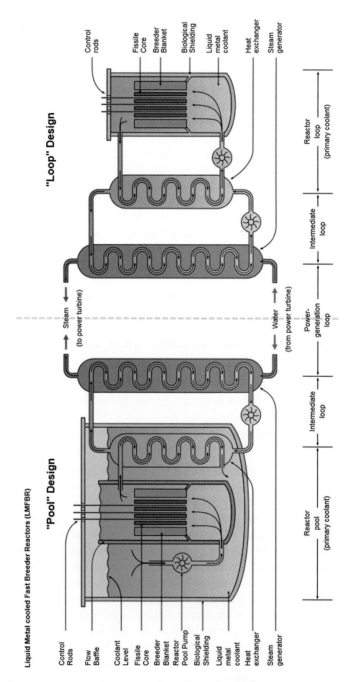

Fig. 4.7 Schematic representation of two types of Liquid-Metal Fast-Breeder Reactor (LMFBR). In this reactor, the fission reaction produces thermal energy to run the turbine while at the same time breeding plutonium fuel for the reactor [12]

FBR can be either in ceramic or metallic form. The ceramic fuels can be classified into oxide, carbide or nitride; the largest operational experience is with oxide fuel, which allows for higher burnup than metallic fuel, because of its physical and chemical characteristics.

Due to the high neutron fluence (i.e. total integrated neutron flux) and temperatures, in a fast neutron reactor stainless steel is used as structural material. Depending upon the arrangement of pumps and heat exchangers in the reactor, fast reactor cores can be either of the loop type or pool type (see Fig. 4.7). In loop type reactors, the primary coolant is circulated through the primary heat exchangers kept outside the reactor vessel. In the pool type, the primary heat exchangers and pumps are immersed in the reactor vessel. Each of these layouts has its own advantages and disadvantages. One main advantage of the pool type reactors is their large capacity of sodium which can act as heat sink during transients.

A significant advantage of a fast reactor is that it makes much better use of the basic uranium fuel, in practice by an estimated factor of 60. Moreover, since a fast breeder reactor breeds new fuel there are subsequent savings in fuel costs since the spent fuel can be reprocessed to recover the usable plutonium. In a thermal reactor, the limited amount of U-235 content in the fuel is utilized as fissile material and the abundant U-238 is largely unused and is disposed of as nuclear waste, while it would be possible to use it as fertile material in a fast reactor. Actually, some countries reprocess the spent fuel, extracting the plutonium and using it together with U-238 to produce MOX fuel to be supplied to thermal reactors (see also Chap. 6). However, this practice is not widespread and moreover the characteristics of the thermal reactor do not make it suitable for breeding (see Chap. 3) so that only a modest improvement in the fuel economy is obtained. With the concept of breeding in fast breeder reactors, the abundant U-238 can be converted into fissile Pu-239 and hence a better fuel utilization is possible in fast reactors. Similarly, in principle the other abundant fertile material Th-232 can also be converted into fissile U-233, with the reaction in Eq. (3.12). Fast reactors using these fuels could provide an energy source for thousands of years.

Furthermore, as the capture to fission ratio falls down at high neutron energies, in a fast reactor the actinide elements have a higher probability of fission than capture. While the fission of a long-lived actinide typically produces two short-lived fission products, the capture keeps producing heavier long-lived actinides. Therefore, fast reactors with low capture to fission ratio can burn long-lived actinides, thereby reducing the problem of long-term storage of radioactive wastes (see Sect. 6.6).

4.5   Generations of Nuclear Reactors

The nuclear reactor technology has been under continuous development since the first commercial exploitation of civil nuclear power in the 1950s. This technological development is presented as a number of broad categories, or *generations*, each

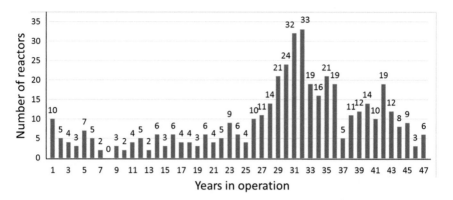

Fig. 4.8 Distribution of operational power reactors by age (as of January 2016). Obtained with data from Ref. [3]. The total number of reactors (441) is higher than that of Table 4.1 (439) because of two new units entered in operation in the last two months of the year 2015

representing a significant technical advance (in terms of either performance, costs, or safety) compared with the previous generation.

The majority of nuclear power plants in operation were designed in the late 1960s and 1970s and their designs are not offered commercially today. Many of the earliest reactors, Generation I, which started commercial operation in the 1950s, had an electrical capacity of 50 MW(e) or smaller. They were power plants of various designs (gas-cooled/graphite moderated, or prototype water cooled and moderated).

Generation II reactors have many representatives in the current fleet of light-water pressurised and boiling water reactors in operation today, which ranges in capacity from 800 to 1500 MW(e). Newer reactor designs increased gradually in power (i.e. size), taking advantage of economies of scale to be competitive. Most of today's nuclear power plants in operation are Generation II reactors originally designed for 30 or 40 year operating lives. About three quarters of all reactors in operation are over 20 years old, and one quarter over 30 years old (see Fig. 4.8). However, based on studies on operating experience and materials behaviour under high temperature and neutron irradiation as well as with major investments in systems, structures and components, lives can be extended. In several countries there are active programs to extend operating lives even 20 years beyond the originally licensed operational life. Periodic safety reviews are undertaken on older plants in line with international safety conventions and principles to ensure that safety margins are maintained (see Chap. 5).

The Generation-III designs are an evolution of current light-water reactor technology with improved performance and extended design lifetimes, as well as more favourable characteristics in case of extreme events such as those associated with core damage (see Chap. 5). The first few Generation III (and Generation III+) advanced reactors were developed in the 1990s with a number of evolutionary designs that offer significant advances in safety and economics, and a number have been built, primarily in East Asia. Some of these reactors were operating in Japan before the Fukushima Daiichi accident in 2011 (see Sect. 5.9), and others are under construction in China, France, Finland, Russia, USA and other countries. A typical example is the EPR—the European Pressurised water Reactor, also called Evolutionary Power Reactor, of which four units are presently under construction in France, Finland and China (two units).

At present, the nuclear research community is looking at a range of innovative reactor designs, so-called Generation IV, that could be commercially deployed starting from 2040. The "Generation IV Technology Roadmap" (see below) gives an overview of the generations of nuclear energy systems. Drawing on earlier designs, today's reactor technology takes into account the following design characteristics:

- Sixty years life;
- Simplified maintenance online or during outages;
- Easier and shorter construction time;
- Inclusion of safety and reliability considerations at the earliest stages of design;
- Modern technologies in digital control and the human-machine interface;
- Safety system design guided by risk assessment;
- Simplicity, by reducing the number of rotating components;
- Increased reliance on passive safety systems (for instance, features relying on gravity, natural convection and not on active controls or on operation intervention to avoid accidents in the event of malfunction, etc.);
- Addition of severe accident mitigating equipment;
- Complete and standardized designs with pre-licensing.

There are a number of international efforts aimed at developing safe and secure innovative nuclear power systems for the long term. Two major international efforts, the Generation IV International Forum (GIF) [13] and the IAEA's International Project on Innovative Nuclear Reactors and Fuel Cycles (INPRO) [14], help participating Member States to evaluate new technology developments and to assess whether nuclear energy would be a viable option and an integral part of their future energy mix.

Six nuclear reactor concepts have been chosen as more promising through expert solicitation and down-selection from a wide range of possible designs. These are the Sodium Fast Reactor (SFR), the Very High Temperature Reactor (VHTR), the Lead Fast Reactor (LFR), the Gas Fast Reactor (GFR), the Super-Critical Water Reactor (SCWR), and the Molten Salt Reactor (MSR). These six concepts are considered to exhibit the greatest potential to show the desired Generation-IV characteristics of increased sustainability,[5] competitive economics, high level of safety, increased

---

[5]In general terms, sustainability is the capacity of our society to maintain itself indefinitely by reducing the human impact on the environment, namely at a level that can be sustained by the planet as a whole. An important aspect of sustainability is the use of natural resources, and indeed sustainable nuclear fission should make better use of them, as discussed in this Chapter.

proliferation[6] resistance (see Sect. 5.11) and, for some designs, the ability to cogenerate high grade heat for use in industrial processes (chemical industry, production of hydrogen or synthetic fuels, etc.).

Generation-IV research covers a broad range of disciplines and areas, and includes work on the fuel cycle (see Sect. 4.6) as well as the reactor components. For example, it aims at developing fast reactors that can also burn the minor actinides recycled from spent fuel. Nowadays, when minor actinides are separated from the spent fuel, they end up in the waste, where they are responsible for much of the heat and radiation produced by the waste in the long term. By recycling the minor actinides back into the reactor, and by careful design of the fuel and operation of the reactor, they can be burnt in the core and transmuted into less radiotoxic and shorter-lived radionuclides (this will be discussed in more details in Sect. 6.6). This is not only an effective way of reducing waste quantities, but the recycling of the minor actinides along with the plutonium also greatly reduces the risk of proliferation because pure bomb-grade plutonium is at no point separated from the other components of the spent fuel. This would make the fuel cycle an extremely unattractive source of nuclear material for an illicit atomic weapons programme (see Sect. 5.10).

4.6 The Nuclear Fuel Cycle

The nuclear fuel cycle is an industrial process involving various activities to produce electricity from the nuclear fuel in nuclear power reactors. The cycle starts with the mining of the nuclear fuel and ends with the disposal of nuclear waste.

The most common raw material for today's nuclear fuel is uranium. It must be processed through a series of steps to produce an efficient fuel for generating electricity. Used fuel also needs to be taken care of for reuse and/or disposal. The nuclear fuel cycle includes the *front end*, i.e. preparation of the fuel, the *service period*, in which fuel is used during reactor operation to generate electricity, and the *back end*, i.e. the safe management of spent nuclear fuel including reprocessing and reuse and disposal.

If the spent fuel is not reprocessed, the fuel cycle is referred to as an *open* or *once-through fuel cycle*; if spent fuel is reprocessed, and partly reused, it is referred to as a *closed nuclear fuel cycle* (see Sect. 4.7).

---

[6]By proliferation in this context it is meant the production of nuclear weapons by more and more countries, in particular those who are not part of the Nuclear Nonproliferation Treaty (see Sect. 5.11), with the increased possibility that the weapons get out of control.

4.6.1 Uranium Mining

Uranium is a common, slightly radioactive metal that occurs naturally in the Earth's crust. It is present in most rocks and soils, in many rivers and in seawater (0.003 parts per million) [15]. Uranium is about 500 times more abundant than gold and about as common as tin.

There are three ways to mine uranium: open pit mines, underground mines and in situ leaching where the uranium is leached directly from the ore. The largest producers of uranium ore are Kazakhstan, Canada and Australia. The concentration of uranium in the ore could range from 0.03 up to 20 %.

4.6.2 Uranium Milling

It is the process through which mined uranium ore is crushed and chemically treated to separate the uranium. The result is the so-called *yellow cake*, a yellow powder of uranium oxide (U_3O_8). In the yellow cake the uranium concentration is raised to more than 80 %.

Milling is generally carried out close to a uranium mine. After milling, the yellow cake concentrate is shipped to a conversion facility.

4.6.3 Conversion

Natural uranium consists primarily of two isotopes, about 99.275 % is U-238 and about 0.720 % is U-235. The fission process by which thermal energy is released in a nuclear reactor mainly involves U-235. Most nuclear power plants require fuel with U-235 concentration increased or, as it is commonly said, *enriched* to a level of 3–5 %. Since enrichment is performed by processing the material in gaseous form, yellow cake is converted to uranium hexafluoride gas (UF_6) at a conversion facility. UF_6 gas is filled into large cylinders where it solidifies. The cylinders are loaded into robust metal containers and shipped to an enrichment plant.

4.6.4 Enrichment

It is the physical process of increasing the isotopic concentration of uranium-235 above the level found in natural uranium. A number of enrichment processes have been demonstrated in the laboratory but only two, the gaseous diffusion process and the centrifuge process, have operated on a commercial scale. Both processes use UF_6 gas as the feed material; fluorine consists of only a single isotope ^{19}F, so that the difference in molecular weight between $^{235}UF_6$ and $^{238}UF_6$ is due only to the small difference in weight of the uranium isotopes. Molecules of UF_6 with U-235 atoms are about one percent lighter than the rest, and this difference in mass is the basis of both processes.

In the centrifuge process, uranium is enriched in U-235 isotope by introducing the gas in fast-spinning cylinders (*centrifuges*), where heavier isotopes are pushed out to the cylinder walls. The method is based on large rotating cylinders with very high rotation speed (50,000 to 70,000 revolutions per minute (rpm)). Because of the effect of the rotation, the heavier $^{238}UF_6$ accumulates towards the chamber outer edge, while the lighter $^{235}UF_6$ concentrates near the centre. By introducing a temperature gradient along the cylinder, it is possible to push $^{235}UF_6$ towards one end and $^{238}UF_6$ towards the opposite end of the cylinder, so that the isotopes can be collected.

Gaseous diffusion (i.e. the rate at which a gas escapes through a pinhole into a vacuum) is an older enrichment technology for uranium enrichment. In this process, UF_6 gas is pumped through a series of porous membranes or diaphragms that allow molecules containing U-235 to pass through more readily than those containing the heavier isotope U-238. Hence the gas which passes through the membrane contains slightly more U-235. At the contrary, the part of gas which did not pass contains slightly less U-235 (it is said that it is *depleted* in U-235). Therefore, by collecting the portion of gas that passed through the membrane one can obtain enriched UF_6 (while the depleted UF_6 is removed). Because of the tiny weight difference between U-235 and U-238 the process must be repeated more than a thousand times in sequential stages (called *cascades*) in order for the end product to be significantly enriched in U-235 (3 to 4 %).

Other enrichment techniques have been studied and developed, for instance based on the use of lasers to provide selective excitation or ionization of the molecules, so that they can be differentiated based on their atomic status. However, these techniques do not have wide commercial application yet.

When considering the costs of production of the nuclear fuel, it turns out that the enrichment step contributes to almost 50 % of the fuel cost, while its impact on the price of the produced electricity is about 5 %. Also, when considering the CO_2 impact in the whole nuclear cycle, the energy needed to run the enrichment process must be taken into account and in the past it used to be the main contributor to it. However, in the case of modern centrifuge facilities, if one assumes that the energy necessary to produce a certain amount of nuclear fuel comes from coal-fired plants, the corresponding CO_2 emissions associated to the enrichment process represent only 0.1 % of the emissions of a coal-fired plant to produce the same amount of energy as the nuclear plant that consumes that amount of nuclear fuel.

4.6.5   Fuel Fabrication

Enriched uranium in the form of hexafluoride (UF_6) cannot be directly used in reactors, as it does not withstand high temperatures or pressures. It is therefore converted into uranium oxide (UO_2) and used to fabricate fuel pellets by pressing

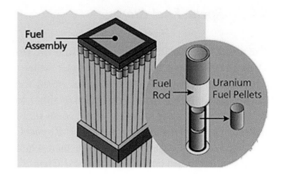

Fig. 4.9 Schematic view of a fuel assembly, with detail on a fuel rod and fuel pellet [15]

UO_2, which is sintered at temperatures of over 1400 °C to achieve high density and stability. The pellets are cylindrical and are typically 8–15 mm in diameter and 10–15 mm high. They are packed in long metallic tubes to form fuel rods, typically up to four metres long. The metal used for the tubes depends on the design of the reactor. Stainless steel was used in the past, but most reactors now use a zirconium alloy (so-called *zircaloy*) which, in addition to being highly corrosion-resistant, has low neutron absorption.

A number of fuel rods are grouped into *fuel assemblies* (also called *fuel bundles*) that are used to build up the core of a power reactor. Figure 4.9 schematically illustrates how the pellets make up a rod and how the rods, in turn, are grouped together to form a fuel assembly.

The specific design and arrangements of the fuel assemblies vary considerably from one reactor type to another. The diameter and length of the rods, the number of rods in an assembly, the presence of empty channels for control rods, neutron sources, test fuel rods or measurement devices, are all different from one case to the other. Just to give an idea, a PWR with 1100 MW(e) power may contain almost 200 assemblies for a total of about 50,000 rods and almost 20 million fuel pellets. A BWR contains between 370 and 800 fuel assemblies, containing about 46,000 fuel rods.

Generally, the fuel loaded into the reactor core remains there for a period of the order of years, which depends on the operating cycle of the specific reactor type. Typically, about once a year, a partial replacement of the used fuel with new one (25–30 %) is necessary. When the fuel replacement is performed, the part, which is not unloaded, is rearranged my moving the fuel assemblies to optimize the core performance.

4.6.6 Electricity Generation

Once the fuel is loaded inside a nuclear reactor, controlled fission can occur. Fission means that the fuel nuclei are split. The splitting releases thermal energy that is used to heat up a coolant fluid (usually water) and produce high pressure steam. The steam turns a turbine connected to a generator, which generates electricity.

4.6.7   Spent Fuel Storage

The spent fuel assemblies removed from the reactor are very hot and radioactive. Therefore the spent fuel is stored under water, which provides both cooling and radiation shielding. After a few years, spent fuel can be transferred to an interim storage facility. This facility can involve either wet storage, where spent fuel is kept in water pools, or dry storage, where spent fuel is kept in casks. Both the heat and radioactivity decrease over time. After 40 years in storage, the fuel's radioactivity will be about a thousand times lower than when it was removed from the reactor. More details are given in Sect. 4.7 and Chap. 6.

4.6.8   Reprocessing

The spent fuel contains uranium (about 96 %), plutonium (about 1 %) and high-level (which means high activity or long lifetime) radioactive waste products (about 3 %) (see Sect. 4.7). The uranium, with less than 1 % fissile U-235, and the plutonium can be reused to make fresh fuel.

Some countries chemically reprocess usable uranium and plutonium to separate them from unusable waste. Recovered uranium from reprocessing can be returned to the conversion plant, converted to UF_6 and subsequently re-enriched in U-235 content. Recovered plutonium, mixed with uranium, can be used to fabricate mixed oxide PuO_2-UO_2 (MOX) nuclear fuel. More details are given in Sect. 4.7 and Chap. 6.

4.6.9   Spent Fuel and High-level Waste Disposal

Spent nuclear fuel or high-level radioactive waste can be safely disposed of deep underground, in stable rock formations such as granite, thus eliminating the risk to people and the environment. The first final disposal facilities will be in operation around the years 2020-2025 (see Sect. 6.4). Waste will be packed in long-lasting containers and buried deep in the geological formations chosen for their favourable stability and geochemistry, including limited water movement. These geological formations have stability over hundreds of millions of years, far longer than the time interval over which the waste can be considered dangerous.

Problem 4.1: Amount of ^{235}U in Uranium Oxide. Calculate how many kilograms of ^{235}U there are in 100 kg of natural uranium oxide U_3O_8.
Solution: The molecule of uranium oxide consists of three atoms of uranium and eight of oxygen. Since the molar masses of U and O atoms are 238 g and 16 g, respectively (we neglect the difference in molar mass between ^{238}U and

^{235}U because of the small percentage of the latter in natural uranium), we have for the U percentage in mass

$$\text{Mass fraction of U} = \frac{3M(U)}{M(U_3O_8)} = \frac{3 \times 238}{3 \times 238 + 8 \times 16} = 0.85.$$

Hence, there are 85 kg of uranium in 100 kg of uranium oxide. Of these 85 kg, only 0.7 % is ^{235}U. Therefore,

$$\text{Mass fraction of U-235} = 0.0072 \times 85 = 0.61.$$

Hence, there are 0.61 kg of U-235 in 100 kg of U_3O_8.

Problem 4.2 Centrifuge in Medical Laboratory. A laboratory centrifuge is a laboratory equipment that spins at high speed liquid samples contained in centrifuge tubes. It works by the sedimentation principle, where the centripetal acceleration is used to separate substances of greater and lesser density dispersed in a liquid. Consider a centrifuge rotor spinning a centrifuge tube, 4 cm long, at $f = 25,000$ rpm. The top of the tube is 5.0 cm from the rotor's central axis, and the bottom of the tube is 9.0 cm from the central axis. Evaluate the acceleration of a particle found at the centre of the tube and compare it with the acceleration of gravity $g = 9.82$ m/s^2.
Solution: A particle traveling in a circle of radius R travels a distance of $2\pi R$ at each revolution. So its speed is

$$v = 2\pi Rf = 2 \times 3.14 \times 0.07 \times \frac{25000}{60} = 183.17 \text{ m/s}.$$

Then the particle acceleration is

$$a = \frac{v^2}{R} = \frac{183.17^2}{0.07} = 5.59 \times 10^5 \text{ m/s}^2.$$

The ratio of this acceleration to g is

$$a = \frac{5.59 \times 10^5}{9.82} = 56,925.$$

Therefore, at 7.0 cm, the centrifugal force on the particle is about fifty-seven thousand g's.
In the centrifuges used to separate the ^{235}U from ^{238}U, the centrifugal force is about 10^4–10^5 stronger than the gravitational force.

Problem 4.3: Graham's Law for Gaseous Diffusion. Gaseous diffusion is based on the Graham's law, which states that the rate of effusion of a gas is inversely proportional to the square root of its molecular mass. Derive this relationship for a monatomic ideal gas using the law of equipartition of energy.

Solution: The law of equipartition of energy states that a system of particles in equilibrium at the absolute temperature T has an average energy of $\frac{1}{2}kT$ associated with each degree of freedom, in which k is the Boltzmann constant. An atom of a monoatomic ideal gas has three degrees of freedom (the three spatial, or position, coordinates of the atom) and will, therefore, have an average total kinetic energy

$$\frac{1}{2}Mv^2 = \frac{3}{2}kT,$$

where M is the molecular mass of the gas. From this equation one immediately gets

$$v = \sqrt{\frac{3kT}{M}},$$

which shows that the rate of diffusion of a gas is clearly inversely proportional to the square root of its molecular mass. Thus, if the molecular weight of one gas is four times that of another, it would diffuse through a porous plug at half the rate of the other. Heavier gases diffuse more slowly.

Problem 4.4: Rate of Diffusion of Uranium Hexafluoride. UF_6 is the only compound of uranium sufficiently volatile to be used in the gaseous diffusion process. Calculate how much faster $^{235}UF_6$ diffuses than $^{238}UF_6$. (The molecular masses of ^{235}U, ^{238}U and ^{18}F are 235.043930 u, 238.050788 u, and 18.998403, respectively).

Solution: From Graham's law, for a given temperature T, the ratio of the average speed of the molecules of $^{235}UF_6$ and $^{238}UF_6$, is inversely proportional to the square root of their molecular masses, M_1 and M_2:

$$\frac{1}{2}M_1v_1^2 = \frac{1}{2}M_2v_2^2 = \frac{3}{2}kT.$$

From which it follows

$$\left\langle \frac{v_1^2}{v_2^2} \right\rangle = \frac{M_2}{M_1}.$$

The molecular masses M_1 and M_2 of $^{235}UF_6$ and $^{238}UF_6$ are

$$M_1 = 235.043930 + 6 \times 18.998403 = 349.034348 \, \text{g/mol},$$
$$M_2 = 238.050788 + 6 \times 18.998403 = 352.041206 \, \text{g/mol}.$$

Then, the ratio between the velocity v_1 and v_2 of the two gasses is:

$$\frac{v_1}{v_2} = \sqrt{\frac{M_2}{M_1}} = \sqrt{\frac{352.041206}{349.034348}} = 1.004298.$$

$^{235}UF_6$ diffuses about 1.0043 times faster than $^{238}UF_6$. Because the velocities of $^{235}UF_6$ and $^{238}UF_6$ are nearly equal, a single pass through a diffusion membrane effects very little separation. Therefore, it is necessary to connect more than a thousand diffusers together in a sequence of stages, using the outputs of the preceding stage as the inputs for the next stage in order to obtain the desired level of enrichment.

Problem 4.5: Identification of an Unknown Gas. An unknown gas diffuses 1.66 times more rapidly than CO_2 (molecular weight 43.99). Identify the unknown gas.

Solution: Using the Graham's law we can determine the molar mass of the unknown gas. Calling r_1 and r_2 the diffusion rate of CO_2 and the unknown gas, and M_1 and M_2 the relevant molar masses, we have:

$$\frac{r_1}{r_2} = \sqrt{\frac{M_2}{M_1}}$$

From which

$$M_2 = M_1 \left(\frac{r_1}{r_2}\right)^2 = \frac{43.99}{1.66^2} = 15.96 \, \text{g/mol}.$$

Then the unknown gas is methane, CH_4, whose molecular mass is 16.031 g/mol.

4.7 Main Fuel Cycles: Open Cycle Versus Closed Fuel Cycle

Figure 4.10 shows the typical isotopic composition of fuel loaded into and discharged at the end of a production cycle (so-called spent fuel) from a Light-Water Reactor. While the numbers will vary slightly depending on the exact enrichment of the fresh fuel, this figure shows the amount of the change in the ^{235}U and ^{238}U content as well as the changes in other isotopes. For convenience, we consider here 1 kg of fresh fuel consisting of ^{238}U and ^{235}U in percentages of 96.7 and 3.3 % (in actual fact, in a typical 1000 MW(e) reactor, the core contains about 75 tonnes of such low-enriched uranium). During operation of the reactor part (2.04 %) of the

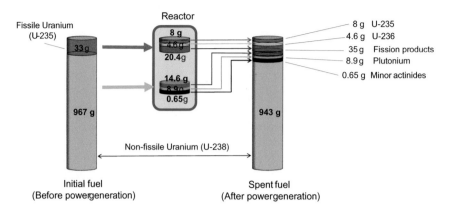

Fig. 4.10 Typical isotopic composition of uranium fuel as loaded to and discharged (after three years of electricity production) from a Light-Water Reactor (LWR). Data from ref. [16]

^{235}U nuclides are fissioned, part (0.46 %) are converted by neutron capture into non-fissile ^{236}U while the remaining part (0.8 %) do not undergo any transformation. About 1 % of the non-fissile ^{238}U isotope is converted into fissile and non-fissile plutonium isotopes by neutron capture. Part of the formed plutonium is also fissioned along the way and so contributes to the energy production. Another part of the plutonium is converted into the minor actinides. The rest of ^{238}U (94.3 %) remains unchanged. Overall, the irradiated fuel contains about 96 % materials that can be reused as fuel; this is uranium that has not undergone transformation processes (94.3 % of ^{238}U and 0.8 % of ^{235}U) and newly produced plutonium (0.89 %).

The final composition of the spent fuel has led to the adoption of two different ways of treating it: the *open cycle* and *closed cycle* (see Fig. 4.11). In the case of the open cycle, known in technical terms as a *once-through cycle*, the uranium is burned once in the reactor and the spent fuel is stored in geological repositories. This method uses less than 2.5 % of the energy content of uranium, which involves the production of large amounts of spent fuel to be stored in safe conditions (see Chap. 6). Both of these problems can be mitigated by recycling spent fuel.

In the closed cycle, the fuel discharged from reactors is chemically reprocessed to recover the uranium and plutonium present that are reused to make fresh fuel. Thus the problem of disposing of materials with high activity arises only for non-reusable materials (about 3 %), the so-called high-activity waste, which include only the fission products and minor actinides (see Sect. 6.6).

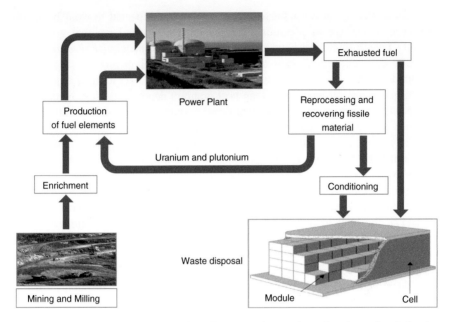

Fig. 4.11 The nuclear fuel cycle comprises all the operations performed on the nuclear fuel, going from the extraction of the fissile material from the ore up to the storage of radioactive waste. In the open cycle (*red arrows*), the uranium is extracted from the ore, turned into fissile material, burned once in the reactor and deposited in a disposal site. In the closed cycle (*green arrows*), the spent fuel is recycled to extract the uranium and plutonium for reuse to produce fresh fuel. In this way, most of the fuel removed from a nuclear reactor is reused thereby reducing the amount of waste to be stored in the final repository

4.8   World Reserves of Nuclear Fuel

4.8.1   Uranium Resources

Uranium is a relatively common element in the crust of the Earth. It is a metal approximately as common as tin or zinc, and it is a constituent of most rocks and even of the sea. Uranium resources are classified by a scheme, based on geological certainty and production costs, developed to combine resource estimates from a number of different countries into harmonised global figures.

Identified resources—which include *reasonably assured resources* (usually abbreviated as RAR), and *inferred resources*—refer to uranium deposits characterized by sufficient direct measurements to conduct pre-feasibility and sometimes feasibility studies. For reasonably assured resources, there is high confidence in estimates of grade and tonnage that are generally compatible with mining decision-making standards. Inferred resources are not defined with such a high degree of confidence and generally require further direct measurements prior to making a decision to mine.

Undiscovered resources refer to resources that are expected to exist based on geological knowledge of previously discovered deposits and regional geological mapping. One distinguishes between *prognosticated* and *speculative resources*. The first refer to resources expected to exist in known uranium provinces, generally supported by some direct evidence. Speculative resources refer to those expected to exist in geological provinces that may host uranium deposits. Both prognosticated and speculative resources require significant amounts of exploration before their existence can be confirmed and grades and tonnages can be defined.

Clearly, the amount of uranium available also depends on the price one wishes to pay for it. The International Atomic Energy Agency (IAEA)[7] and the Nuclear Energy Agency (NEA)[8] are probably two of the most competent sources of information on this topic. They publish every year a document called the IAEA-NEA Red Book "Uranium: Resources, Production and Demand". According to the Red Book 2014 [17], the total identified resources (reasonably assured and inferred) as of January 1, 2013 amounted to 5,902,900 tonnes of uranium metal (tU) in the category up to 130 US\$/kgU. In the highest cost category up to 260 US\$/kgU total identified resources amounted to 7,635,200 tU. At the 2012 level of uranium demand (61,980 tU/a), identified resources appear sufficient for over 120 years of supply for the global nuclear power fleet.

Moreover, an additional 119,100 tU of resources have been identified by the IAEA/NEA as resources reported by companies that are not yet included in national resource totals. Total undiscovered resources (both prognosticated and speculative) as of January 1, 2013 amounted to 7,697,700 tU. Another potential source of uranium, of much larger size, is the seawater, where the uranium is present in solution with a concentration of approximately 0.003 parts per million [15]. It is estimated that a quantity of uranium equal to about 4 billion tonnes may be available in this form, which is however more difficult and costly to retrieve.

Figure 4.12 shows the global distribution of the established reserves of uranium in the world [17]. It can be seen that Australia hosts a substantial part (about 29 %) of the world's uranium, Kazakhstan 12 %, Russian Federation 9 %, Canada 8 % and Niger 7 %. The top five uranium-rich countries account for nearly 65 % of the world's uranium resources and over 75 % of the world's total uranium output. These stocks are widely distributed geographically and are in accessible locations. This geopolitical distribution of resources and production guarantees a relatively high security of supply, at variance with the case of oil.

World's nuclear weapons stockpiles are also an important source of nuclear fuel. Indeed, weapons typically contain uranium enriched to over 90 % in U-235 (much

---

[7]The International Atomic Energy Agency (IAEA) is the autonomous, intergovernmental organisation set up in 1957 by the United Nations. Its mission is to promote the safe, secure and peaceful use of nuclear technologies and to inhibit its use for any military purpose.

[8]The Nuclear Energy Agency (NEA) is a specialised agency within the Organisation for Economic Co-operation and Development (OECD). Its mission is to assist its member countries in maintaining and further developing the scientific, technological and legal bases required for a safe, environmentally friendly and economical use of nuclear energy for peaceful purposes.

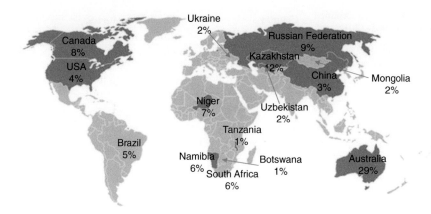

Fig. 4.12 World's proven uranium resources as of January 1, 2013. Data from Ref. [17]

more than reactor fuel). Some weapons instead contain Pu-239, which is a component of the MOX fuel. Based on specific disarmament treaties with the USA (starting from 1987), countries from the former Soviet Union (USSR) agreed to reduce their stockpiles by about 80 % and to offer their weapons-grade uranium to produce ordinary nuclear fuel. Between 1999 and 2013, a total of 500 tonnes of military highly-enriched uranium from the former USSR have been diluted with enriched uranium relatively low in U-235 content, to produce over 14,000 tonnes of uranium at standard enrichment to be used as civil fuel, a quantity sufficient to supply the world's reactor fleet for 2.5 years [18].

4.8.2   Thorium Resources

Currently uranium is the only nuclear fission fuel of natural origin. An alternative fuel may be obtained from thorium, provided that specific reactors are designed for this purpose.

As a natural element, thorium is half as abundant as lead and is three times more abundant than uranium in Earth's crust [19]. Essentially 100 % of thorium is the isotope ^{232}Th, which undergoes α-decay with a half-life of 14 billion years (more than three times the Earth's age).

When bombarding thorium with neutrons, it can be transformed into uranium-233 (see Eq. (3.12)), a fissile isotope that can sustain a chain reaction. In the transmutation process the amount of waste produced is relatively limited. Since the starting isotope has mass number $A = 232$, while most of the uranium has mass number 238, thorium-based fuels produce less plutonium than those based on uranium and, very important, with isotopic composition unsuitable for the manufacture of explosive devices. Moreover, also the production of other heavier, long-lived actinides like americium and curium is suppressed. For all these reasons,

Table 4.6 World's identified thorium resources [20]

Country	Thorium resources	
	[kt]	(%)
India	846	13.3
Brazil	632	9.9
Australia	595	9.4
USA	595	9.4
Egypt	380	6.0
Turkey	374	5.9
Venezuela	300	4.7
Canada	172	2.7
Russia	155	2.4
South Africa	148	2.3
China	100	1.6
Norway	87	1.4
Greenland	86	1.4
Finland	60	0.9
Sweden	50	0.8
Kazakhstan	50	0.8
Other	1725	27.1
World total	**6355**	**100**

the exploitation of thorium as nuclear fuel, or in other words the so-called *thorium fuel cycle*, has been considered with great interest for many years and is back in vogue in newly designed reactors having the purpose to test the concepts at the basis of its use. However, there are technical aspects in the fuel cycle that make it difficult to adopt the thorium fuel cycle in a cost-effective-manner and that require sizeable investments to promote the necessary R&D. Currently, China and India are the countries dedicating most efforts to studying this possibility, with some support from the USA.

As of January 2013, the total identified resources (reasonably assured and inferred), with an associated extraction cost of 80 US$/kgTh or less, amounted to 6,355,300 tonnes of thorium [20]. As it is seen in Table 4.6, they are rather widespread geographically, occurring on all continents, with the top eleven countries accounting for about 68 % of the total resources.

4.8.3   Uranium Demand

Uranium demand is expected to continue to rise for the foreseeable future from the about 62,000 tU/a required to fuel the 437 commercial nuclear reactors in operation in 2012 to between 72,000 tU/a and 122,000 tU/a by 2035. At that date, also taking into account changes in policies in several countries following the Fukushima

Daiichi nuclear power plant accident occurred in 2011 (see Sect. 5.9), world nuclear capacity is projected to grow to between about 400 GW(e) net in the low demand scenario and about 678 GW(e) net in the high demand scenario from the 2012 value of 372 GW(e) net. This represents increases of about 8 and 82 %, respectively [17, 21].

Nuclear capacity projections vary considerably from region to region. The East Asia region is projected to experience the largest increase, which, by the year 2035, could result in the installation of between 55 GW(e) and 125 GW(e) of new capacity in the low and high scenarios, respectively, representing increases of more than 65 and 150 % over the capacity at the end of 2012. Nuclear capacity in non-EU member countries on the European continent is also projected to increase significantly, with additions of between 20 and 45 GW(e) of projected capacity by 2035 (increases of about 50 % and 110 % respectively). Other regions projected to experience significant nuclear capacity growth include the Middle East, Central and Southern Asia and South-East Asia, with more modest growth projected in Africa and the Central and South American regions. For North America, nuclear generating capacity in 2035 is projected to either decrease by almost 30 % in the low scenario or increase by over 20 % in the high scenario. In the European Union the outlook is similar, with nuclear capacity in 2035 either projected to decrease by 45 % in the low scenario or increase by 15 % in the high scenario [17].

These projections are subject to great uncertainty due to the Fukushima Daiichi accident, since Japan has not yet determined the role that nuclear power will play in its future generation mix and China did not report official targets for nuclear power capacity beyond 2020 [17]. Key factors influencing future nuclear energy capacity include projected baseload electricity demand, the economic competitiveness of nuclear power plants, as well as funding arrangements for such capital-intensive projects, the cost of fuel for other electricity generating technologies, non-proliferation concerns, proposed waste management strategies and public acceptance of nuclear energy. The latter is a particularly important factor in some countries after the Fukushima Daiichi accident. Concerns about longer-term security of fossil fuel supply and the extent to which nuclear energy is seen to be beneficial in meeting greenhouse gas reduction targets and enhancing security of energy supply could contribute to an even greater projected growth in uranium demand.

At the consumption rate of the end of 2012, the about 5.9 million tonnes of low cost identified resources of uranium would last for about 95 years, while the total of about 15 million tonnes from identified low and high cost reserves plus the undiscovered ones would be enough for about 240 years.

In fact, it seems reasonable to expect that both the rate of consumption and the expansion of available resources will change, along with their more efficient use. In particular, the introduction of fast reactors of Generation IV in a fuel closed cycle would increase the efficiency of uranium burnup by a factor of 60 compared to the open cycle currently adopted for many LWRs. Even in the event of having to wait until the end of the current century for their full operation, fast reactors would allow to increase the uranium resources remaining at that date (about 10 million tonnes) for the above factor 60, which, in terms of duration, would mean almost 10,000 years, at the consumption rate of the end of 2012. Moreover, all this disregarding uranium seawater and the reserves of thorium.

Therefore, uranium is an abundant resource, more of which could be made available for the nuclear industry by expanding the exploration activities and opening new mines. Some risks for the security of supply might come not from the limited resources or political instability, but rather from possible delays in moving from discovery of new sources to production, particularly if demand will grow rapidly.

Problem 4.6: Annual consumption of nuclear fuel. Calculate how many kilograms of ^{235}U per year are burned in a nuclear reactor operating for 365 days at a thermal power $P = 3400$ MW(th).

Solution: In one year (that is in approximately 3.15×10^7 s) the reactor will produce the energy

$$E = \left(3.4 \times 10^9 \, \text{W}\right) \times \left(3.15 \times 10^7 \, \text{s}\right) = 10.7 \times 10^{16} \, \text{J}.$$

Recalling that the fission of one fissile nucleus yields about 200 MeV = 3.2×10^{-11} J of energy, in order to produce the energy E one has to fission

$$N = \frac{10.7 \times 10^{16}}{3.2 \times 10^{-11}} = 3.35 \times 10^{27} \, \text{nuclei}^{235}U.$$

Calling $N_A = 6.02 \times 10^{23}$ the Avogadro's number and assuming that the atomic weight of ^{235}U is 235, this corresponds to a mass

$$M = N\frac{235}{N_A} = \frac{3.35 \times 10^{27} \times 0.235}{6.02 \times 10^{23}} = 1308 \, \text{kg}.$$

Problem 4.7: Total Mass of Uranium Used in a Power Plant. Calculate how much ^{235}U is needed for the lifetime of a nuclear power plant that produces 1000 MW of electricity. Consider that the lifetime of the plant is 40 years and that its efficiency to convert heat into electricity is 33 %. Assume that the efficiency of conversion of nuclear energy into heat is 100 % and that the power plant operates at 85 % load factor.

Solution: Taking into account that $1 \, a \approx 3.15 \times 10^7$ s, and that the power plant produces 1000 MW(e) with a load factor of 85 %, the amount E_{el} of electric energy produced in one year is

$$E_{el} = \left(10^9 \, \text{J/s}\right) \times \left(3.15 \times 10^7 \, \text{s}\right) \times 0.85 \approx 2.68 \times 10^{16} \, \text{J per year}.$$

At 33 % efficiency of thermal-to-electric conversion, the thermal energy E_{th} required to produce this amount of electricity is:

$$E_{th} = \frac{E_{el}}{0.33} = \frac{2.68 \times 10^{16}}{0.33} \approx 8.12 \times 10^{16} \, \text{J per year.}$$

Then, over a lifetime of 40 years, the quantity of thermal energy required for the plant will be:

$$\left(8.12 \times 10^{16} \, \text{J}\right) \times (40 \, \text{a}) \approx 3.25 \times 10^{18} \, \text{J.}$$

In the Problem 3.3, we calculated that the fission of 1 kg of U-235 releases approximately 8.2×10^{13} J of energy. Then to operate the power plant it is needed a mass of fissile uranium

$$M = \frac{3.25 \times 10^{18}}{8.2 \times 10^{13}} = 3.96 \times 10^4 \, \text{kg} \approx 40,000 \, \text{kg of } ^{235}\text{U.}$$

So about 40 tonnes of fissile uranium are needed for a typical nuclear reactor over a lifetime of 40 years.

Problems

4.1 Evaluate the most probable velocity of thermal neutrons at a temperature of (a) 20 °C and (b) 260 °C.
[*Ans.*: (a) 2692 m/s; (b) 3620 m/s]

4.2 If equal amounts of helium (molecular mass M_1 = 4.003 u) and argon (molecular mass M_2 = 39.948 u) are placed in a porous container and are allowed to escape, which gas will escape faster and how much faster?
[*Ans.*: helium will escape 3.16 times faster]

4.3 Assume you have a sample of hydrogen gas containing H_2, HD, and D_2 that you want to separate into pure components. What are the various ratios of relative rates of diffusion? (For simplicity, assume the atomic masses to be 2, 3 and 4, respectively)
[*Ans.*: H_2 diffuses 1.414 times faster than D_2, and 1.225 faster than HD; HD diffuses 1.155 times faster than D_2]

4.4 A separation plant extracts heavy-water from natural water with an efficiency of 70 %. Knowing that deuterium is present in natural water in the ratio of 1 to 6420 with respect to hydrogen, how much water must be processed in order to produce 1.0 litre of heavy-water?
[*Ans.*: about 9170 litres of water]

4.5   In 2011 the average consumption of electricity per person in the USA was about 3.3×10^{11} J per year. Knowing that the fission of 1 kg of ^{235}U releases an energy of about 8.2×10^{13} J, calculate how many kilograms of ^{235}U must be burned to produce that energy.
[*Ans.*: 0.004 kg]

4.6   A power station produces about 40 TWh of electrical energy in one year by fission of uranium. (This energy was roughly the electrical consumption of Greater London in the year 2012). The overall efficiency of the process that converts the thermal energy released in the fission to electricity is 40 %. (a) How much energy is produced each second in the reactors of the power station? (b) What is the total mass change due to fission in the reactors of the power station each second?
[*Ans.*: (a) 1.14×10^{10} W(th); (b) about 1.3×10^{-7} kg]

4.7   With the reference to the thermal energy calculated in the previous problem, and assuming that on an average the fission of a single nucleus of ^{235}U releases about 200 MeV of energy, estimate: (a) How many nuclei of ^{235}U must undergo fission in each second to produce the calculated thermal energy. (b) What was the mass of these atoms before they underwent fission? (The atomic mass of ^{235}U is 235,04394 u).
[*Ans.*: (a) 1.78×10^{20} nuclei/s, (b) 6.95×10^{-5} kg]

4.8   Estimate the initial mass of ^{235}U necessary to fuel for one year a reactor that has an electric-power generation capacity of 650 MW(e) and a thermal-to-electricity efficiency of 40 %. (Assume that each fission of ^{235}U nuclei releases an energy of about 200 MeV).
[*Ans.*: 625 kg]

4.9   A pressurised water reactor (PWR), which burns natural uranium (with 0.7 % of ^{235}U), produces 1.0 GW electric power, with a thermal-to-electric efficiency 0.40. The fuel rods stay in the reactor for about 3 years and are then removed to allow for reprocessing. (a) Calculate the mass of ^{235}U required in the core for a 3-year cycle. (b) Estimate the total mass of both the uranium isotopes required in the core for a 3-year cycle.
[*Ans.*: (a) about 2884 kg; (b) 412,000 kg]

4.10 A thermal nuclear reactor contains a total uranium mass of 70 tonnes with 3.5 % ^{235}U enrichment. It develops an electric power $P = 900$ MW(e) with a 35 % thermal-to-electricity efficiency. Calculate (a) the number of ^{235}U nuclei fissioning every second and (b) the equivalent mass of enriched uranium burned per day. (c) Assuming the plant works at constant full power, how long can it run before changing the fuel? (assume the change to occur when all ^{235}U has been consumed).
[*Ans.*: (a) about 8.0×10^{19} (nuclei ^{235}U)/s; (b) about 77 kg/d of uranium; (c) about 2.5 a]

References

1. OECD-NEA Nuclear Energy Today, 2nd edn. (2012), ISBN 978-92-64-99204-7. NEA Report No. 6885
2. IAEA, Nuclear Power Reactors in the World, 2015 Edition. Online http://www-pub.iaea.org/books/IAEABooks/10903/Nuclear-Power-Reactors-in-the-World-2015-Edition.
3. IAEA Power reactor information system (PRIS), http://www.iaea.org.pris
4. http://www.world-nuclear.org/Press-and-Events/Briefings/Restart-of-Sendai-1/
5. IAEA-PRIS, MSC, *2015: The World nuclear Industry Status Report 2015*, by M. Schneider, A. Froggatt, J. Hazemann, T. Katsuta. M.V. Ramana, S. Thomas, J. Porritt. Paris, London, July 2015
6. WNA, *Nuclear Power in the World Today*, http://www.world-nuclear.org/info/current-and-future-generationnuclear-power-in-the-world-today/
7. J. Hermans, *Energy Survival Guide* (Leiden University Press/BetaText, 2011), p. 165
8. W. Moomaw, P. Burgherr, G. Heath, M. Lenzen, J. Nyboer, A. Verbruggen, *Renewable Energy Sources and Climate Change Mitigation*, Special Report of the Intergovernmental Panel On Climate Change (IPCC), p. 19, and Annex II: Methodology (2001), p. 190. ISBN 978-92-9169-131-9. Online: https://www.ipcc.ch/pdf/special-reports/srren/SRREN_FD_SPM_final.pdf.
9. M. Ricotti, *Nuclear Energy: Basics, Present, Future,* Lecture Notes of the Course 1 "New strategies for energy generation, Conversion and storage" of the Joint EPS-SIF International School on Energy, Varenna, Lake Como, 30 July–4 Aug (2012). ISSN 2282-4928 and ISBN 978-88-7438-079-4
10. WNA, *Power Reactors—Characteristics*. 2010 WNA Pocket Guide, World Nuclear Association, July 2010
11. http://www.britannica.com/science/sodium
12. https://upload.wikimedia.org/wikipedia/commons/4/46/LMFBR_schematics2.svg   http://creativecommons.org/licenses/by-sa/3.0/
13. GenIV International Forum, https://www.gen-4.org/gif/jcms/c_9260/Public
14. INPRO, https://www.iaea.org/inpro/
15. http://world-nuclear.org/information-library/nuclear-fuel-cycle/uranium-resources/supply-of-uranium.aspx
16. http://www.stormsmith.nl/i11.html
17. OECD-NEA&IAEA, "*Uranium 2014: Resources, Production and Demand*" (*The Redbook*). Online: https://www.iaea.org/OurWork/ST/NE/NEFW/Technical-Areas/NFC/uranium-production-cycle-redbook.html
18. WNA, http://www.world-nuclear.org/info/nuclear-fuel-cycle/uranium-resources/military-warheads-as-a-source-of-nuclear-fuel/
19. http://www.britannica.com/science/thorium
20. WNA, http://www.world-nuclear.org/info/current-and-future-generation/thorium/
21. WNA, http://www.world.nuclear.org/info/Nuclear-Fuel_Cycle/Uranium-Resources/Supply-of-Uranium

Chapter 5
Nuclear Safety and Security

Besides being a mandatory part of general policies to protect the public and the environment, nuclear safety and security are also critical for the public acceptance of nuclear power plants. This chapter reviews the safety and security aspects involved in the use of energy from nuclear fission, as well as the regulatory system and the international bodies that have been established to address them. The ten most relevant accidents occurred at nuclear installations are briefly described, together with the corresponding lessons learned, that—as for any other complex technology—have greatly contributed to the progressive improvements in safety. Finally, the last Section reports on the security measures and regulations in force to prevent and detect any misuse of nuclear technologies.

5.1 Nuclear Safety Regulations

The main objective of nuclear safety is the achievement of proper operating conditions of the nuclear installations, the prevention of accidents and the mitigation of their consequences, in order to guarantee the protection of the workers, the public and the environment from undue radiation hazards from authorised activities. A radiation hazard, or radiological hazard, is the health and environmental risk associated with the exposure to ionising radiation. Similarly, an accident with radiological consequences is one where unwanted exposure to radiation of workers and/or the public and/or the environment can occur.

Since the first application of nuclear energy, safety issues have always had a primary importance in the design, construction and operation of nuclear installations and in the whole nuclear fuel cycle, because of the enormous potential hazard resulting from the accumulation of large amounts of radioactive products in the fuel. Therefore, the use of radioactive materials and the processes by which they are

© Springer International Publishing Switzerland 2016   189
E. De Sanctis et al., *Energy from Nuclear Fission*, Undergraduate Lecture
Notes in Physics, DOI 10.1007/978-3-319-30651-3_5

produced are strictly regulated to ensure nuclear safety. In particular, regulations establish maximum exposure levels for both workers at nuclear installations and the public. On the other hand, efforts are always made to stay well below maximum levels allowed.

Each country is responsible for the safety of the nuclear power plants within its borders. Governments are responsible for establishing an independent "Regulatory Authority" and for enacting legislation to regulate their own nuclear activities, including the management of spent fuel and radioactive waste. A regulatory authority is a national public body that is responsible for overseeing nuclear activities, making sure that the principles of safety and security are properly followed and that rules established by the law are obeyed.

Ensuring nuclear safety also requires the availability of suitably qualified staff, the establishment of an effective safety culture in the workforce, the funding of research on operational and safety issues and an appropriate focus on security. The work of national regulators covers all these aspects. National regulatory authorities adhere to the principles of "good regulation", which include independence, technical competency, transparency, efficiency, clarity and reliability. A fundamental aspect regarding the regulatory authority is that it must be independent, i.e. it must not have any connection with the regulated entities (e.g. companies designing, building or operating nuclear power plants), nor any conflict of interest regarding the regulated matter.

A few specialised international organisations have been founded, with the mission to help nations develop nuclear energy for peaceful purposes, and to promote the development of relevant safety and regulatory concepts in the use of nuclear technologies and the adoption of good practices. Among these organisations, the International Atomic Energy Agency (IAEA) and the Nuclear Energy Agency (NEA) are especially authoritative and independent. The IAEA is the autonomous, intergovernmental organisation set up in 1957 by the United Nations; as of the end of September 2015, it had 165 member states. The NEA is the specialised agency established in 1958 by the Organisation for Economic Co-operation and Development (OECD); it brings together 31 countries from North America, Europe and the Asia-Pacific region. A co-operation agreement is in force between these two Agencies.

A Convention on Nuclear Safety was drawn up by the IAEA between 1992 and 1994 through dedicated expert meetings. The Convention was approved in 1994, entered into force in 1996 and was recently amended on February 9, 2015 by the Vienna Declaration on Nuclear Safety [1]. The objective of the Convention is to achieve and maintain a high standard of nuclear safety all over the world. Such objective must be realized through the implementation of effective measures in nuclear installations, in order to prevent potential radiological hazards and accidents with radiological consequences. As a practical framework for achieving its objective, the Convention sets specific safety benchmarks that the individual States must subscribe. This means in turn that each State should enact national legislation that receives the subscribed benchmarks, so that the high safety level goals defined by the benchmarks become legally binding for that State and the companies operating nuclear power plants on its territory.

A detailed description of the nuclear safety principles and obligations to which all subscribers should adhere are found in a specific IAEA document [2]. They range from the legislative and regulatory framework, to several technical safety aspects such as those related to siting, design, construction, operation of nuclear installations, to quality assurance and emergency preparedness.

As of September 2009, there were 79 signatories to the IAEA Convention, including all countries with nuclear power plants in operation. The national regulatory bodies of the latter countries are entrusted with the responsibility to verify the fulfilment of all safety regulations. The licence for construction of a nuclear plant is subject to a positive assessment by the national regulatory authority on the characteristics of the site and the plant, and also the license for operation of the plant is subject to an equally positive outcome of various tests. In addition, during the operation, the authority periodically carries out routine inspections. As a result, the plant operation can even be stopped for some time, pending the implementation of modifications that, in some cases, can be substantial. National regulatory authorities also contribute to informing citizens about the hazards of nuclear power and radiation as well as about safety provisions and measures adopted in nuclear installations.

5.2 Safety and Radiation Protection Objectives

All nuclear facilities are sited, designed, constructed and operated in accordance with strict quality and safety standards, whose goal is to ensure that the operation of plants does not contribute significantly to individual or societal health risks. Therefore, reactor safety regards the prevention of radiation-related damage to the plant workers, the public and the environment from the operation of commercial nuclear reactors. The same principle applies to any nuclear installation.

There are three so-called fundamental safety objectives: (i) the general nuclear safety objective, (ii) the radiation protection objective, and (iii) the technical safety objective. These objectives require that nuclear installations are designed and operated so as to keep all sources of radiation exposure under strict technical and administrative control. More specifically, the general nuclear safety objective requires that all sources of possible malfunctions and accidents be taken into account, so as to minimize the probability of major failures and accidents. The radiation protection objective requires that all scenarios have been examined and all possible measures undertaken, in order to minimize radiation exposure of the plant workers, the public and the environment. The technical safety objective requires that all scenarios have been examined and measures undertaken in order to protect the plant equipment.

In order to achieve the above mentioned three safety objectives, when designing a nuclear power plant a comprehensive safety analysis is carried out, with the purpose to identify all sources of exposure and to evaluate radiation doses that could be received by workers at the installation and the public, as well as potential effects on the environment. The safety analysis takes into consideration: (i) all planned normal operational modes of the plant, (ii) the plant performance in anticipated operational occurrences, (iii) the design basis accidents, and (iv) the beyond design basis accidents, including severe accidents.

Anticipated operational occurrences include events of rare occurrence that can be expected to happen during normal operation of the plant. They include for example a loss of power to all cooling pumps, loss of power from the main external grid, etc.

A *Design Basis Accident* (DBA) [3], is a postulated accidental scenario that a nuclear facility must be designed and built to withstand without fatal consequences for the systems, structures, and components necessary to ensure safety. Design Basis Accidents typically include external occurrences like earthquakes, floods, severe weather events, etc., as well as internal ones like fires, uncontrolled reactivity (and power) surges, etc. Accident conditions more severe than DBA are called *Beyond Design Basis Accidents* (BDBA); BDBA involving significant core degradation are called severe accidents.

The goal of the safety analysis is to establish and assess all possible scenarios, from normal operation of the reactor to a series of accidents, that may occur due to both internal and external causes, and that can range from rare but anticipated to occur a few times during the plant lifetime (e.g. severe weather), to rare and not necessarily occurring during the plant lifetime but possible (e.g. earthquakes). On the basis of this analysis, the robustness of the engineering design in withstanding postulated initiating events and accidents can be established, the effectiveness of the safety systems and safety procedures can be demonstrated, and requirements for emergency response can be established.

Although measures are taken to control radiation exposure in all operational states to levels *As Low As Reasonably Achievable* (so-called ALARA principle) and to minimize the likelihood of an accident that could lead to a loss of control of the source of radiation, there is a residual probability that such an accident may happen. Accidental measures are therefore taken to ensure that the radiological consequences are mitigated. Such measures include: on-site accident management procedures established by the operating organization as well as engineered safety systems, and possibly off-site intervention measures established by appropriate authorities in order to mitigate radiation exposure to the public and the environment if an accident has occurred.

The engineered safety systems include the equipment and components necessary to monitor the nuclear facility operation and to ensure the three fundamental safety functions: (i) shut down the reactor, (ii) provide cooling of the fuel, and (iii) in the event of an accident, ensure that radioactive material is securely contained inside the reactor building.

On-site accident management procedures include documentation and training of the personnel for them to be able to handle accidental scenarios like a fire in the plant or a radiological emergency. Off-site intervention measures will include, wherever necessary, communication with civil authorities to undertake appropriate actions in case of hazard for the public or for the environment and in case that outside help be needed (e.g. to fight a major fire).

5.3 The Concept of Defence-in-Depth

The basic design approach followed all over the world for assuring nuclear power plant safety is called *defence-in-depth*, with multiple safety systems supplementing the natural features of the reactor core [4]. Defence-in-depth consists of a hierarchical deployment of different levels of equipment and procedures in order to maintain the effectiveness of physical barriers placed between radioactive materials and workers, the public or the environment. It must be effective in all instances examined in the safety analysis: in normal operation, in anticipated operational occurrences and, for some barriers, in case of accidents at the plant. Defence-in-depth is implemented both in design and operational aspects to provide a graded protection against a wide variety of transients and accidents, including equipment failures and human errors within the plant as well as events initiated outside the plant.

Defence-in-depth is generally structured in five levels, as shown in Fig. 5.1 and in Table 5.1. Should one level fail, the subsequent level comes into play.

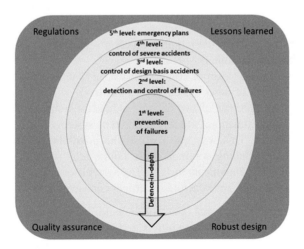

Fig. 5.1 Diagram illustrating the philosophy and the elements of the defence-in-depth concept [4]. It consists of a set of actions, items of equipment or procedures, classified in levels, the prime aim of each of which is to prevent degradation liable to lead to the next level and to mitigate the consequences of failure of the previous level. The defence-in-depth concept now comprises five levels. Should one level fail, the subsequent level comes into play. At the four corners are reported the principles guiding the implementation and improvement of the defence-in-depth

Table 5.1 Levels of defence-in-depth

Levels of defence-in-depth	Objective	Essential means
Level 1	Prevention of abnormal operation and failures	Conservative design and high quality in construction and operation
Level 2	Control of abnormal operation and detection of failures	Control, limiting and protection systems and other surveillance features
Level 3	Control of accidents within the design basis	Engineered safety features and accident procedures
Level 4	Control of severe plant conditions, including prevention of accident progression and mitigation of the consequences of severe accidents	Appropriate containment, complementary measures and accident management
Level 5	Mitigation of radiological consequences of significant releases of radioactive materials	On-site and off-site emergency response

The aim of the first level of defence is to prevent deviations from normal operation, and to prevent system failures. This leads to the requirement that the plant be soundly and conservatively designed, constructed, maintained and operated in accordance with appropriate quality assurance requirements and engineering practices, such as the application of

- redundancy, i.e. the use of multiple systems capable of performing the same specific function, e.g. cooling the core. In other words the provision of alternative—identical or diverse—structures, systems and components, so that any one of them can perform the required function regardless of the state of operation or failure of any other;
- independence, i.e. a system or component whose function is unaffected by the operation or failure of other systems/components;
- diversity, i.e. the presence of two or more redundant systems or components to perform the same function, where the different systems or components have different attributes. This way one reduces the possibility of failure due to a common cause, including common cause failure (e.g. when one event causes several systems or components to fail). The systems or components may differ in technology, design, manufacture, electrical connections, software, etc.

If the first level fails, abnormal operation is controlled or failures are detected by the second level of protection whose aim is to detect and intercept deviations from normal operational states in order to prevent anticipated operational occurrences from escalating to accident conditions. For instance, a failure in a cooling pump will be detected and quickly repaired, while the other pumps (redundancy) keep circulating the coolant and maintaining the core temperature at safe levels.

Should the second level fail, the third level ensures that safety functions are further performed by activating specific safety systems and other safety features.

Specifically, the third level of defence assumes that, although very unlikely, escalation of certain anticipated operational occurrences may not be controlled by a preceding level of defence, and a more serious event may develop. These unlikely events are anticipated in the design basis for the plant (recall the DBAs mentioned above), and inherent safety features, fail-safe designs, and additional equipment and procedures are provided to control their consequences and to restore the plant to a safety condition following such events.

The aim of the fourth level of defence is to address severe accidents in which the design basis may be exceeded and to ensure that radioactive releases are kept as low as achievable. The most important objective of this level is to guarantee the confinement function by limiting accident progression through accident management, so as to prevent or mitigate severe accident conditions with external release of radioactive materials. This is primarily achieved by appropriate reactor containment. The Three Miles Island accident (see Sect. 5.9) in which the nuclear fuel melted was a severe accident but there was not any significant release of radioactivity into the environment because the containment building remained intact. By contrast, in the Chernobyl accident (see Sect. 5.9) a large amount of radioactivity was released into the environment because the plant did not have a suitable containment structure as found in the other Generation II or current Generation III nuclear power plants.

The fifth and final level of defence, i.e. the off-site emergency plan, aims at the mitigation of the radiological consequences of significant releases of radioactive materials from the reactor containment. This requires the provision of an adequately equipped emergency control centre and plans for the on-site and off-site emergency response, and more generally it requires to elaborate a series of short- and long-term protective actions that need to be taken.

Table 5.1 summarises the objectives of each level of protection of the defence-in-depth concept and the primary means of achieving them.

A relevant aspect of the implementation of defence-in-depth is the provision in the reactor design of a series of physical barriers to confine the radioactive material at specified locations. Their specific design may vary depending on the activity of the material and on the possible deviations from normal operation that could result in the failure of some barriers. For most of the reactors, the barriers confining the radioactive products are typically: (i) the pellet material containing the fuel (called the *matrix*), (ii) the fuel cladding, (iii) the boundary of the reactor cooling system, (iv) the containment system (see next Section).

The plant workers, the public and the environment are protected primarily by means of these barriers, which may serve operational and safety purposes (e.g. in a water-cooled reactor, the pressure vessel keeps the cooling fluid at high pressure and it also provides a barrier in case of failure of the fuel matrix and the cladding) or safety purposes only (e.g. the containment building). The defence-in-depth concept applies to the protection of the integrity of the barriers against internal and external events that may jeopardize them. Situations in which there is a potential breach of one or more barriers necessitate special attention.

5.4   Reactor Safety

In a nuclear reactor, fresh fuel is only weakly radioactive if it is uranium-based, as it contains the two isotopes ^{235}U and ^{238}U, which decay with very long half-life, or only moderately radioactive if it is plutonium-based (MOX fuel, see Sect. 4.6). Moreover, the metal cladding surrounding it protects the pellets, preventing any unwanted dispersion of radioactive material. However, as we have seen in Sect. 3.12, a nuclear reactor produces various radioactive substances: fission products (like caesium, strontium, iodine, xenon etc.), heavy elements produced by transmutation (neptunium, plutonium, americium, curium, etc.), fluids that get contaminated by contact with the fuel rods, and structural materials made radioactive mainly by neutron capture. It is therefore necessary to exercise maximum care in the reactor design, operation and decommissioning to avoid component failures and accidents than may lead to an unwanted and uncontrolled release of radioactive materials into the external environment.

Ensuring that a reactor can be operated safely is one of the most important goals of reactor designers and operators. In order to minimize the potential hazard to the public from the radioactive materials contained within an installation, a number of principles and provisions have been developed and incorporated into the design and operation of nuclear power plants. Collectively, these principles are summarised in the golden rule of Reactor Safety, which can be stated as: "there is a minimum risk to the public and the environment from reactor operation, provided that at all times: (a) the reactor power is controlled, (b) the heat generated in the core is removed, (c) the radioactivity is contained."

In terms of reactor generations (see Sect. 4.5), commercial units of Generation II already incorporated a number of active safety components, where active means that their functioning depends on an external input such as actuation, mechanical movement or supply of power. Some new designs in Generation III and Generation III+ also incorporate passive safety systems that are activated by natural physical phenomena, like gravity.

5.4.1   Control of the Reactor

As outlined in Sect. 3.10, many physical phenomena can change the reactor status and affect the reactivity. If the reactivity decreases, the changes will have to be compensated such as to bring it back to the critical level to avoid a loss of reactor power. The danger can come from any situation where the reactor power rises in an uncontrolled way, typically because of an accidental scenario where some control component has failed. Therefore, as a main safety feature, all the reactors are designed to be inherently safe, which means that during any unwanted rise of reactor power the overall reactivity feedback will be negative which will result in a reduction of power. This feature prevents the uncontrolled rise of power.

As described in Sect. 3.10, the reactor power control during normal operation of the plant is performed by the control rods, which are special bars containing neutron absorbers, like boron or cadmium, that are suitably withdrawn from or inserted into the core to regulate the power. For completely shutting down the reactor, the control rods are fully inserted into the core. Moreover, all reactors have to be equipped with an emergency shutdown system in case the chain reaction has to be stopped immediately because of a change in the reactor status that can pose an immediate danger (e.g. failure of some essential component). Such sudden stop is called a *SCRAM*. Historically, it is reported that Enrico Fermi invented this expression during the startup of the first experimental reactor at the University of Chicago. The expression would have meant "Safety Control Rod Axe Man", as there would have been a single safety rod hanging from a rope and there would have been a person in charge of quickly cutting the rope, letting the rod drop down, in case of emergency. However, there may have been other equivalent meanings of this acronym.

In Pressurised Water Reactors (PWRs), the shutdown systems are designed to be "fail safe", as the control rods are clutched to electromagnets at the top of the reactor vessel. In case of electrical power failure, they simply drop into the core by gravity. In Boiling Water Reactors (BWRs), the control rods are inserted into the core from below and therefore no spontaneous fall by gravity is possible. In this case, the shutdown system is based on a hydraulic high-pressure mechanism that pushes the rods into the core. In addition, for all reactors a secondary procedure for emergency shutdown is provided, e.g. by the injection of neutron-absorbing liquids (like boron-rich water), to ensure long-term reactor shutdown. More advanced reactor designs also adopt additional passive shutdown systems. In Sect. 5.9, we briefly describe the Chernobyl accident, where an uncontrolled rise of power occurred, mainly due to issues with the reactivity control and with the design of the control rods.

5.4.2 Removal of Heat Generated in the Core

During normal operation, thermal energy is generated in the core due to nuclear fission and also to a minor extent due to nuclear decay. Therefore, the core needs to be cooled-down in all operational conditions. Removal of thermal energy is performed by pumping a coolant, i.e. a suitable fluid that circulates through the core (see schematic figures of reactor installations in Chap. 4). Normally, two or more cooling circuits are installed for redundancy.

Even when the reactor is in shutdown state, a certain amount of residual thermal energy is generated due to the decay of fission products (so-called *decay heat*). This decay heat reduces with time because radioactive nuclei decay. At shutdown time of a reactor that has operated for a while, thereby reaching an equilibrium between production and decay of fission products, the decay heat amounts to almost 6.5 % of the nominal thermal power of the reactor. After 1 h, it decreases to about 1.5 %; after one day, it will fall to about 0.4 %; and after one week it will be only 0.2 %.

For a reactor of 3 GW thermal power, this means that at shutdown time the residual decay heat is about 200 MW, while after one day is about 12 MW, and after one week is about 6 MW. The latter is still a considerable amount of power, which is why the core has to be cooled for long time after the shutdown. For the same reason, spent fuel rods removed from the core are kept in special water pools to cool down before being further processed. A reactor is defined to be in *cold shutdown* when not only the chain reaction is stopped by full insertion of the control rods, but also the coolant within the circulation system is at atmospheric pressure and at a temperature below 100 °C.

The redundancy in the number of cooling systems helps in decay heat removal in case of failure of one loop. Furthermore, all water-cooled reactors are equipped with an *Emergency Core Cooling System* (ECCS), which is independent from normal cooling circuits. The emergency core cooling system allows for thermal energy removal from the core even if there is a leak in primary coolant circuits.

As an additional emergency safety feature, the cooling system pumps are provided with a backup power supply based on diesel generators and battery banks, which supply power in case of a failure of the main external power grid. Finally, in some advanced reactors the cooling circuits are designed in such a way that the decay heat can be passively removed by natural convection, i.e. by spontaneous circulation of the fluid between the hot core and the colder heat exchangers. Hence, in these particular designs, even when all sources of power are lost, decay heat removal from the core is ensured.

Failure in cooling the core after shutdown can produce severe damage to the core by melting the fuel pellets and cracking the fuel rods and can also give rise to dangerous hydrogen production if water steam at very high temperature interacts with the zirconium present in the fuel cladding. Removal of decay heat is therefore a crucial aspect in designing a reactor and implementing safety measures, as the Three Miles Island accident and, recently and more dramatically, the Fukushima Daiichi accident have stressed (see the corresponding discussion in Sect. 5.9).

5.4.3   Containing the Radioactivity

Radioactive materials, and in particular fission products, are produced in the core of the reactor when the fission and capture processes occur. Most of these radioactive products remain within the fuel itself. To prevent their release into the environment, at least four successive barriers are in place (see Fig. 5.2).

The first barrier is the fuel pellet matrix, where the fission products are generated. The second barrier is the cladding which contains the fuel. The third barrier is the leak-tight boundary of the reactor cooling system. The fourth barrier is the containment building. The latter, in some designs, comprises two barriers, where the first can be for example a reinforced steel barrier with approximately spherical shape, surrounding the pressure vessel, and the second is a reinforced concrete building surrounding the first containment (the concrete building, however, does

Fig. 5.2 Typical barriers confining radioactive materials [5]

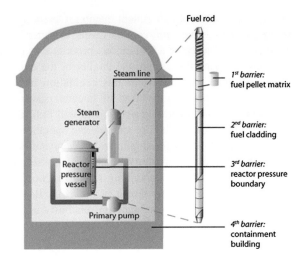

not guarantee a complete radioactive tightness). Since the four barriers are independent, failure of all four barriers at the same time is highly unlikely.

Another reason why the radiological impact of the Chernobyl accident was extremely high was that such particular type of reactor had no containment building (see Sect. 5.9).

5.5   Safety in the Design, Operation and Decommissioning

The design of a nuclear power plant shall ensure that the systems, structures and components important to safety have the appropriate characteristics, specifications and material composition so that the safety functions can be performed and the plant can operate safely with the necessary reliability for the full duration of its design life, with accident prevention and protection of site personnel, the public and the environment as prime objectives.

The design shall also ensure that the requirements of the operating organization are met, (e.g. providing the required energy production and load factor, the latter being the number of hours in a year during which the plant is fully productive), and providing that due attention is given to the foreseen capabilities of the personnel who will eventually operate the plant. The design organization must supply adequate documentation to ensure safe operation and maintenance of the plant and to allow for subsequent plant modifications to be made, and must recommend practices for incorporating them into the plant administrative and operational procedures, taking into account operational limits and conditions.

Feedback from operating experience helps to ensure and enhance safety in operating plants and to prevent severe accidents, in particular by using lessons learned from accident precursors (i.e. specific events which may lead to an accident by starting a chain of subsequent events). Operating experience also indicates the significance of events for various levels of defence-in-depth. The evaluation of operating experience is a continuous process that involves checking the assumptions made during the design, the quality of the construction and the adequacy of plant operation. The results of this evaluation have significantly influenced the design of the current generation of nuclear power plants as well as backfilling measures[1] taken in operating plants, and will influence the design of future plants.

The continuous availability and reliable operation of the engineered systems are key elements of defence-in-depth, and their operation is regularly tested. Design of these systems must ensure that the failure of any single safety component would not cause the loss of the corresponding safety function.

In the design of nuclear power plants, some degree of conservatism is applied, meaning that previous experience on reliability and performance of materials and systems is highly valued. However, also a certain extent of pessimism must be applied if risks are to be correctly assessed. For example, in the accident at the Fukushima Daiichi plant in 2011 (see Sect. 5.9), from the preliminary analysis of the event it seems that the risk of flooding from an exceptionally powerful tsunami was underestimated by not properly considering extremely rare events. Measurements of the physical and chemical properties of all materials used is also an important aspect in design choices.

The safety analysis relies both on the so-called *Deterministic Safety Approach* (DSA) and the so-called *Probabilistic Safety Assessment* (PSA). The first considers a series of design basis accidents (DBAs) without considering their probability, thereby assuming a pessimistic worst failure of a single safety component and analysing the response of the plant. The second introduces probabilities for equipment failures and human errors, based in part on previous data from real plant operation. PSA is a comprehensive, structured approach to identifying failure scenarios, providing a conceptual and mathematical tool for deriving numerical estimates of risk. Three levels of probabilistic safety assessment are generally recognized. Level 1 comprises the assessment of plant failures leading to determination of the frequency of core damage. Level 2 includes the assessment of containment response, leading, together with Level 1 results, to the determination of frequencies of failure of the containment and release into the environment of a given percentage of the reactor core's inventory of radionuclides. Level 3 includes the assessment of off-site consequences, leading, together with the results of Level 2 analysis, to estimates of public risks. All PSAs are used for making decisions on plant modifications, new designs, operator training and inspections.

It is evident that, as in all industrial fields, to guarantee maximum safety, in addition to all safety analyses of any kind, it is the people who have to carry out all

---

[1]Backfilling measures are modifications of an existing plant, implemented to take into account the feedback from the operating experience and to correct weaknesses pointed out in the corresponding evaluation process.

activities necessary to support the safety of a plant. Therefore, it is of fundamental importance to promote: (i) a good safety culture, (ii) the principles of good management, (iii) a continuous surveillance and periodic safety analyses, (iv) the use of lessons learned from actual experience, (v) the communication of errors and issues found.

The use of high-quality components is also essential for reliable operation. Therefore, a crucial component of nuclear safety is quality assurance, which is carried out by using specific codes and standards for equipment and components. Quality assurance also includes tests and inspections whose purpose is to guarantee that well-established technologies are adopted. The primary responsibility for the implementation of the quality assurance and control programmes rests on the plant operator, while national regulatory authorities oversee the process.

All assessments regarding safety are contained in specific documents called "safety analysis reports", which are submitted to the regulatory authorities for review and approval. Approval of the safety documentation is an essential step in the licensing of a nuclear installation, i.e. the permissions by the authorities to first build and then to operate the plant. Such documentation then becomes the reference to safely operate the facility.

Before operation, a nuclear power plant must first undergo a commissioning stage. This includes both nuclear and non-nuclear testing. Systems and components of the plant are made operational, tested and verified to be in accordance with the design and to have met the required performance criteria. Subsequently, the overall behaviour of the plant is tested and, if necessary, corrective actions are applied. Overall checks must also be conducted after major maintenance operations, replacement of components and system upgrades.

Often, safety assessments are performed repeatedly throughout the lifetime of a plant, together with self-assessments conducted by the organisation operating the installation or by independent peers. Self-assessments and independent safety peer reviews ensure that plants can continue to operate without a safety breach. In addition, international peer reviews are conducted by the IAEA and by the World Association of Nuclear Operators (WANO),[2] while global reviews are conducted by signatories of the Convention on Nuclear Safety, mentioned in Sect. 5.1.

Safety regulations and requirements apply also to decommissioning, which is the end of the operating lifetime of a nuclear power plant, when it is retired from service. Decommissioning implies a sequence of operations that start with unloading the spent fuel from the reactor, typically followed by an intermediate stage where the fuel elements are left to cool in special water pools. Then all the components of the plant, starting from the reactor pressure vessel to steam units,

---

[2]The World Association of Nuclear Operators (http://www.wano.info/en-gb) is a non-profit organization that gathers every company and country in the world hosting operating commercial nuclear power plants, with the mission to achieve the highest possible standards of nuclear safety.

turbines, pipes, facilities, buildings, etc. have to be dismantled and all materials have to be separated according to their radioactivity levels (see Chap. 6). Depending on the legislation of the state hosting the plant, some materials may be released as non-radioactive materials and either disposed of as normal waste or recycled without limits. The radioactive materials on the contrary could be reused in the nuclear industry if their radioactivity is very low. Otherwise, they will have to undergo special processes like purifying, drying, compacting, etc., then will have to be sealed in suitable packages (e.g. metal drums or concrete matrices) and finally disposed of in appropriate licensed repositories where the materials are kept under surveillance.

5.6 Responsibility for Safety and Regulation

The responsibility for the safety of the nuclear power plants rests on the organization operating the plant, while the responsibility for issuing specific laws to regulate the sector, in particular the establishment of an independent regulatory authority, rests on the governments. Specifically, the tasks of the Regulatory Authority are:

- to provide safety requirements and guidelines,
- to verify compliance with regulations by also verifying that plant designs comply with safety requirements,
- to issue licences for siting, construction and operation of the plant,
- to perform inspections and to monitor and review the safety performance of the plant operators,
- to enforce safety regulations by imposing the necessary corrective measures.

However, the prime responsibility for safety rests on the operator that holds the license to operate the plant, for the whole lifetime of the facility, while other parties like plant designers, vendors, manufacturers, etc. are responsible for their professional activities regarding safety, according to the corresponding contracts in place.

Clearly, it is essential that there be a full separation between the regulatory organisation and all other parties involved, so that every decision process regarding nuclear energy be free from conflicts of interest and lobbying pressures. As mentioned above, international co-operation through organisations such as the IAEA, the NEA and, in Europe, the European Commission is also a vital aspect in promoting and supporting safety at any level.

5.7 Types of Nuclear Accidents and Accident Management

As discussed in the previous Sections, the concern about safety is a primary aspect in all stages of nuclear energy production. Indeed, the application and evolution of basic safety principles and regulations has made it possible to operate several commercial reactors for many years without causing hazard for people and the environment. However, accidents have happened (see Sect. 5.9) and no analysis nor planning can reduce their probability to zero. Therefore, it is also fundamental to analyse accidental scenarios and to develop action plans directed toward minimizing their consequences.

As mentioned above, a Design Basis Accident is defined as accident conditions against which a facility is designed according to established design criteria, and for which the damage to the fuel and the release of radioactive material are kept within authorized limits.

A Beyond Design Basis Accident comprises accident conditions more severe than a design basis accident, and may or may not involve significant core degradation (like core melting due to insufficient removal of residual decay heat). Accident conditions more severe than a design basis accident and involving significant core degradation are termed *severe accidents*. More precisely, severe accidents are event sequences that may arise in a nuclear power plant owing to multiple failures of safety systems leading to significant core degradation and loss of the integrity of many or all of the barriers whose purpose is to confine the radioactive material.

Consideration of beyond design basis accidents at nuclear power plants is an essential component of the defence-in-depth approach used in ensuring nuclear safety. Clearly, the probability of occurrence of a beyond design basis accident is very low, but such an accident may lead to significant consequences resulting from the degradation of nuclear fuel.

Accident management is essential to ensure effective defence-in-depth at the fourth level (see Sect. 5.4). It comprises a set of actions to be undertaken during the evolution of a beyond design basis accident. Its goals are to prevent the escalation of the event into a severe accident, to mitigate the consequences of a severe accident, and to achieve a long-term safe stable state.

5.8 Previous Experience and Safety Record

The cumulative experience of commercial reactor operation in the whole world amounts to nearly 16,000 reactor-years[3] [6]. The information and lessons learnt from this vast experience are reported in databases, documents by international organisations, journals and conferences.

---

[3]A reactor-year is one year of operation of a single reactor. The number of reactor-years is obtained by summing the number of operating reactors all over the world year by year. For instance, if 10

Concerning databases, a relevant one is the Power Reactor Information System (PRIS) [7], developed and maintained by the IAEA for over four decades. PRIS is a comprehensive database focusing on nuclear power plants worldwide, containing information on power reactors in operation, under construction, or being decommissioned. The database covers: (i) reactor specification data (status, location, operator, owner, suppliers, milestone dates) and technical design characteristics, (ii) performance data including energy production and energy loss data, outage and operational event information. The PRIS public web site [7] provides information on PRIS and global nuclear power reactor statistics to the general public. A second database providing information on events of safety significance is the IAEA/NEA International Reporting System for Operating Experience (IRS) [8]. IRS also contains lessons learned which assist in reducing recurrence of events at other nuclear plants; however it is only open to technical experts in the field. Another major database is the WANO comprehensive system [9], which is however only open to its members.

It is a precise requirement to the plant operators from regulatory authorities to report on all events significant from the point of view of safety, such as human errors, equipment failures, accidents and many others.

As one possible indicator of the operational safety performance one can use the number of unplanned automatic reactor shutdowns (SCRAMs) for one year of operation (approximately 7000 h). This number has decreased markedly from the value 1.8 in the early 1990s to about 1.0 in the years 2003–2004 and down to about 0.6 in the years 2009–2014, which can be interpreted as an improvement in plant operation (see Fig. 5.3 that shows the mean SCRAMs occurred in the years 2003–2014).

Overall, by using these and many other indicators, it is possible to assert that the safety record of worldwide plants has been improving over the years. However, three severe accidents at commercial nuclear power plants have spoiled this good overall picture: at Three Mile Island in the United States in 1979, at Chernobyl in Ukraine in 1986 and at Fukushima Daiichi plant in Japan in 2011. In both the last two accidents, significant releases of radioactivity into the environment have occurred.

(Footnote 3 continued)

reactors have each run for one year, the accumulated experience will be 10 reactor-years. If 100 reactors have each run for 10 years, the accumulated experience will be 1000 reactor-years. If the fact that the reactors are not all equal is neglected, it can be considered equivalent to a single reactor operating for 1000 years. To be more accurate, one should calculate the number of reactor-years for each single reactor class to obtain information on the performance of each individual class.

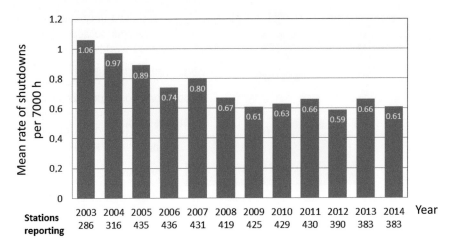

Fig. 5.3 Mean rate of SCRAMs: the number of automatic and manual shutdowns that occurred per 7000 h of operation in the years 2003–2014 [7]

5.8.1 The International Nuclear Event Scale (INES)

The degree of severity of nuclear and radiological accidents is now established through a numerical scale, the International Nuclear Event Scale (INES), elaborated by a group of international experts convened jointly by the IAEA and NEA in 1990. Initially the scale was applied to classify events at nuclear power plants, then it was extended and adapted to enable it to be applied to all installations associated with the civil nuclear industry. More recently, it has been extended and adapted further to meet the growing need for communication of the significance of all events associated with the use, storage and transport of radioactive material and radiation sources. Indeed the primary purpose of the INES is to facilitate communication and understanding between the technical community, the media and the public on the safety significance of events. Its purpose is to keep the public as well as nuclear authorities accurately informed on the occurrence and consequences of reported events.

The INES (see Fig. 5.4 and Table 5.2) uses a numerical rating to explain the significance of events associated with sources of ionising radiation. Events are rated at seven levels: Levels 1–3 are *incidents* or *anomalies* and Levels 4–7 *accidents*. The scale is designed such that the severity of an event is approximately ten times greater for each increase in level of the scale. These levels consider three areas of impact: (i) people and the environment, (ii) radiological barriers and control, and (iii) defence-in-depth. Events without safety significance are rated as *below scale/level 0*. Events that have no safety relevance with respect to radiation or nuclear safety are not rated on the scale.

Table 5.2 presents the INES and safety criteria with more details. The second column of the Table refers to the effects of unwanted releases of radioactivity and

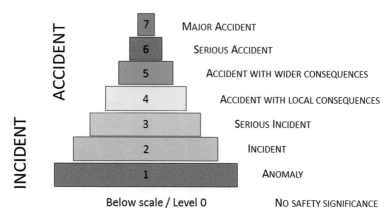

Fig. 5.4 The International Nuclear Event Scale (INES)

Table 5.2 Basic structure of the International Nuclear Event Scale (INES)

INES level and description	Criteria or safety attributes			Examples of accidents at power plants
	People and the environment	Radiological barriers and controls at facilities	Defence-in-depth	
Level 7: MAJOR ACCIDENT	Major release of radioactive material. Widespread health and environmental effects. Implementation of planned and extended countermeasures.			**Chernobyl**, Ukraine, 1986 (fuel meltdown and fire); **Fukushima Daiichi**. 2011 (fuel damage, radiation release, and evacuation)
Level 6: SERIOUS ACCIDENT	Significant release of radioactive material Implementation of planned			**Kyshtym**, Russia, 1957 (military reprocessing plant criticality)

(continued)

Table 5.2 (continued)

INES level and description	Criteria or safety attributes			Examples of accidents at power plants
	People and the environment	Radiological barriers and controls at facilities	Defence-in-depth	
	countermeasures likely.			
Level 5: ACCIDENT WITH WIDER CONSEQUENCES	Limited release of radioactive material. Implementation of some planned countermeasures likely. Several deaths from radiation.	Severe damage to reactor core. Release of large quantities of radioactive material within an installation. High probability of significant public exposure.		**Three Mile Island**, USA, 1979 (fuel melting) **Windscale**, UK, 1957 (military)
Level 4: ACCIDENT WITH LOCAL CONSEQUENCES	Minor release of radioactive material. Implementation of planned countermeasures other than local food controls unlikely. At least one death from radiation.	Fuel melt or damage to fuel resulting in more than 0.1 % release of radionuclides from the core. Release of significant quantities of radioactive material within an installation. High probability of significant public exposure.		**Saint-Laurent**, France, 1969 (fuel rupture) and 1980 (graphite overheating) **Tokai-Mura**, Japan, 1999 (criticality in fuel plant for an experimental reactor)
Level 3: SERIOUS INCIDENT	Exposure in excess of ten times the statutory annual limit for workers. Non-lethal deterministic health effects (e.g. burns) from radiation.	Exposure rates of more than 1 Sv/h in an operating area. Severe contamination in an area not expected by design. Low probability of significant public exposure.	Near accident at a nuclear power plant with no safety provisions remaining.	**Vandellòs**, Spain, 1989 (turbine fire) **Davis-Besse**, USA, 2002 (severe corrosion) **Paks**, Hungary 2003 (fuel damage)
Level 2: INCIDENT	Exposure of a member of the	Radiation levels in an operating	Significant failures in safety provisions but	

(continued)

Table 5.2 (continued)

INES level and description	Criteria or safety attributes			Examples of accidents at power plants
	People and the environment	Radiological barriers and controls at facilities	Defence-in-depth	
	public in excess of 10 mSv. Exposure of a worker in excess of the statutory annual limits.	area of more than 50 mSv/h. Significant contamination in an area not expected by design.	with no actual consequences.	
Level 1: ANOMALY	None	None	Overexposure of a member of the public in excess of statutory limits. Minor problems with safety components with significant defence-in-depth remaining.	
Level 0: DEVIATION. Below scale	None	None	None	

Detailed definitions are provided in the INES User's Manual [10]. Only the parts referring to nuclear installations are reported here (the original table also describes events relative to radioactive sources). The indicated impacts on people and the environment are not necessarily present all at the same time (for instance in the Three Mile Island accident, which was classified at level 5, there were no ascertained deaths).

radioactive substances on the installation workers, the public and the environment. The third column refers to the local effects of the various events on the status of the equipment and the radiological conditions of the installations. The fourth column refers to the level of integrity of the various barriers that form the defence-in-depth. The fifth column lists examples of rated accidents at nuclear installations.

According to this scale, of the three above mentioned most severe accidents in commercial nuclear power plant history, the Three Mile Island accident was rated 5 (accident with wider consequences based on the impact on the installation), while both the Chernobyl and Fukushima Daiichi accidents were rated 7 (major accident), as they caused major releases of radioactive material. The third most severe nuclear accident ever recorded occurred at the Kyshtym military reprocessing plant in Russia, when it was part of the former Soviet Union. It was rated 6 (serious accident) as it led to a widespread external release of radioactive material.

5.9 The Nuclear Accidents

Perfect safety does not exist. Despite the adoption of rigorous safety measures and regulations, nuclear plants, like any other industrial activity, cannot be entirely risk-free. Potential sources of problems include human errors and external events having a greater impact than anticipated. Accidents may happen and, in fact, a number of them have occurred. Analyses of the causes of their occurrence have provided valuable knowledge and lesson learned, leading to progressive improvements in safety.

In the following, we give a short description, in chronological order of occurrence, of the ten most relevant accidents occurred to nuclear installations. The lessons learned from them are emphasized since they have led to substantial improvements of the nuclear power stations all over the world. The accidents are listed in the fifth column of Table 5.2, and correspond to rating 4 or higher on the INES. The four accidents with off-site consequences, the three at commercial nuclear power plants of Three Mile Island, Chernobyl, and Fukushima, and that at the military plant of Kyshtym are discussed in more detail.

5.9.1 Kyshtym (1957), Russia

A serious nuclear contamination accident occurred on September 29, 1957 at the Mayak nuclear energy complex located at 150 km southeast of Yekaterinburg, in Russia. The complex was a plutonium production site for nuclear weapons and nuclear fuel reprocessing plant in the former Soviet Union. The accident was named after Kyshtym, the nearest known town (the closed small city built around the Mayak plant, named Ozyorsk, was not marked on the map for secrecy reasons).

The cooling system failure in one of the tanks containing liquid radioactive waste caused an explosion estimated to have a force equivalent to about 70–100 t of trinitrotoluene explosive. There were no immediate casualties because of the explosion, but about 7.4×10^{17} Bq of radioactive material were released into the atmosphere. Most of this contamination settled out near the site of the accident, but a plume containing 74×10^{15} Bq of radionuclides was deposited as dry fallout over an area some 30–50 km in width and some 300 km in length stretching north-northeast of the nuclear facility. The area is usually referred to as the East-Ural Radioactive Trace (EURT).

Because of the secrecy surrounding the plant, the populations of affected areas were not initially informed of the accident. A week later (on October 6), an operation for evacuating about 10,000 people from the affected area started, still without giving an explanation for the reasons of the evacuation. Vague reports of a catastrophic accident causing radioactive fallout over the Soviet Union territory and many neighbouring States appeared in the western press in April 1958, but it was only in 1976 that the Soviet emigrant Z.A. Medvedev first revealed some facts about the disaster [11]. To reduce the spread of radioactive contamination after the accident, contaminated soil was excavated and stockpiled in fenced enclosures. The Soviet government in 1968 disguised the EURT area by creating the East-Ural Nature Reserve, which prohibited any unauthorised access to the affected area. In 2002, the level of radiation in Ozyorsk itself was claimed to be safe for humans, but the area of EURT was still heavily contaminated with radioactivity.

The event was classified at Level 6 of the INES, as a serious accident that required implementation of the planned countermeasures, making it the third most serious nuclear accident ever recorded.

5.9.2   Windscale Pile (1957), UK

The military nuclear reactor facility and plutonium-production plant of Windscale (now Sellafield) in the county of Cumberland (now part of Cumbria), in north-western England, consisted of two gas-cooled nuclear reactors built as part of the British atomic bomb project.

The accident occurred on October 8, 1957, when the core of Unit 1 reactor caught fire during a routine maintenance check, igniting about 11 tonnes of uranium. It took workers three days to put out the fire. In the meantime, radiation escaped through the chimney, contaminated the surrounding area, and reached as far as mainland Europe. No one was evacuated from the area, but there was a worry that milk might be dangerously contaminated. As a precautionary measure, milk from about 500 km^2 of nearby countryside was diluted and destroyed for about a month. The contaminated Windscale reactor was subsequently sealed until the late 1980s, when its decommissioning begun. Completion of the decommissioning is not expected until at least 2037.

The Windscale fire was the worst nuclear accident in the UK's history. It was rated 5 in the INES.

5.9.3   Three Mile Island (1979), USA

The reported information about the accident is mainly based on publications by NEA [12] and WNA [13].

The partial core meltdown at the Unit 2 of the Three Mile Island site, near Harrisburg, Pennsylvania in the United States, is the most serious nuclear accident in USA history, although it resulted in only small radioactive releases.

The site hosted a nuclear power plant with two reactors, Units 1 and 2, whose construction was completed between 1974 and 1978. Unit 2 (called TMI-2) was a 2568 MW(th) pressurised water reactor (PWR). The accident initiated on March 28, 1979 with an automatic reactor shutdown caused by a relatively minor malfunction in the cooling circuits. A valve left closed by mistake caused a pressure increase in the primary coolant circuit, which in turn caused the opening of a relief valve. After the automatic shutdown, the relief valve had to close automatically, but instead remained open due to a malfunction. Because of an incorrect interpretation of the instrument readings, the operators thought that the valve was closed and stopped the automatic high-pressure emergency cooling system, so that the water level in the core decreased, leaving the fuel assemblies uncovered. This caused the core to overheat and, eventually, a significant portion of the fuel melted and flowed into the lower part of the core and lower reactor vessel head.

Due to the damage to the fuel elements in the core and the fact that there was an open valve, volatile fission products, mainly noble gases, iodine and caesium isotopes left the pressure vessel and accumulated within the containment building, along with a considerable quantity of hydrogen gas.[4] Such hydrogen subsequently underwent a slow burn-up process without compromising the stability of the containment structure. About 5 % of the radioactive noble gases and a very small amount of iodine-131 were released in a controlled way outside of the containment building, however without exposing the population to a dose above background levels. There were no fatalities, injuries or adverse health effects due to the accident on either the workers or the public. During the nuclear materials release, pre-school and pregnant or nursing moms were evacuated within 5 miles of the plant and people were cautioned to stay indoors within a 10 mile radius.

Because of the concern caused by the accident about the possibility of radiation-induced health effects in the area surrounding the plant, the Pennsylvania Department of Health maintained for 18 years a registry of more than 30,000 people who lived within five miles of the plant at the time of the accident. This registry was discontinued in mid-1997, without any evidence of unusual health trends in the area.

The accident was rated 5 on the INES because of the severe damage to the reactor core and the substantial release of radioactivity inside the installation. Investigations following the accident showed that it was due to mechanical failure along with operator confusion, and led to a new focus on the human factors in nuclear safety. As immediate measures, controls and instrumentation were improved significantly and operator training was overhauled. In the longer term, important design upgrades were implemented, so that today, some Generation III reactors (and also some Generation II reactors) feature systems such as core catchers and hydrogen recombiners. A core catcher is a special pool beneath the pressure vessel where the melted core, called *corium*, can be stopped and contained in case it was able to penetrate the pressure vessel and fall down outside of it. A hydrogen recombiner is a catalyst based on a porous material, where hydrogen and oxygen atoms react to form water in a controlled way.

---

[4]Hydrogen is generated when water steam reaches such a high temperature that it can chemically react with the zirconium contained in the fuel cladding according to the following chemical reaction:

$$Zr + 2H_2O \rightarrow ZrO_2 + 2H_2.$$

5.9.4 Saint-Laurent (1980), France

The Saint-Laurent nuclear power plant is located in the commune of Saint-Laurent-Nouan in Loir-et-Cher (France) on the Loire River—28 km downstream of Blois and 30 km upstream of Orléans. At the time of the accident, two reactors (A1 and A2) existed at the site, which were brought into service in 1969 and 1971 and were retired in April 1990 and June 1992.

On March 13, 1980, a failure in the cooling system caused the melting of one channel of fuel in the A2 reactor. No radioactive material was released outside the site. The accident was classified as 4 on the INES. It was the worst nuclear accident in France.

Today the plant includes two operating pressurised gas-cooled, graphite moderated reactors each of 1690 MW(th) power, which began operation in 1983.

5.9.5 Chernobyl (1986), Ukraine

The reported information about the accident is mainly based on publications by NEA [12] and WNA [14].

The world's worst nuclear accident occurred in 1986 at Chernobyl nuclear power plant lying about 130 km north of Kiev, Ukraine, and about 20 km south of the border with Belarus. The area surrounding the plant was woodland with a low population density (within a 30 km radius of the power plant, the total population was between 115,000 and 135,000).

At the time of the accident, four 1000 MW(e) RBMK reactors were in service at the plant with a further two under construction. Each reactor system consisted of two identical reactors back-to-back, rated at 500 MW(e), using 2 % U-235 enriched uranium dioxide fuel, a graphite moderator, and water as coolant (see Fig. 4.5). Among the characteristics of RBMK reactors built at the time, a relevant one was that the specific reactivity coefficients related to the physics of the neutrons and to the thermal behaviour of the core made the system intrinsically unstable at low power, so that it was necessary to rely on control rods to ensure stability. Another peculiar characteristic of this plant was the absence of the containment building, which is the fourth and last barrier against the release of radioactivity foreseen in the golden rule of reactor safety (see Fig. 5.2). The reason for this was that the reactor had been designed to produce weapons-grade plutonium, which required a very frequent replacement of the fuel, which is not feasible with a containment like the one in use in all other civil reactor types. Designers and operators knew these characteristics.

The accidental sequence started on April 26, 1986, at unit 4, when the reactor operators started a test after disconnecting the reactor automatic safety system purposefully. The goal of the test was to determine if a turbine-generator would still provide power to some reactor cooling pumps, while it ran down after its steam supply was removed. The test had a legitimate purpose but was not properly planned: workers did not implement adequate safety precautions or alert operators about the risks of the electrical test, and the test was performed at low reactor power, an unstable condition as mentioned above. Moreover, the control rods had a drawback in their design that caused an increase in reactivity instead of a decrease when they were only partially inserted into the core. This positive reactivity feedback caused a sudden power surge, with power increasing up to a factor hundred before the control rods could be fully inserted. Therefore, the accident was caused by a sudden change in reactivity. The rapid power rise caused an over-heating which eventually destroyed the fuel and, by contact of the hot fuel with water in the cooling channels, caused a steam explosion that ruptured the core containment. Therefore, the accident quickly hit all first three barriers (fuel pellet matrix, fuel cladding and reactor pressure boundary) against the release of radioactivity. As mentioned above, the fourth barrier, the containment building, was absent. Due to the extremely high temperature, the water steam reacted with the zirconium cladding producing a large quantity of hydrogen gas that exploded and destroyed the reactor building. At the same time, the graphite from the moderator caught fire, so that the explosions threw around burning graphite debris that in turn started other fires.

The disintegration of the fuel elements along with the explosions and the graphite fire—that continued for days—caused the release of large amounts of solid and gaseous radioactive materials, which, given the absence of a containment building, were thrown high up in the outside atmosphere. In the following days, several radioisotopes reached and deposited over several European countries, where they were detected by nuclear instrumentation. In Western Europe, contamination was first detected by operators of a Sweden nuclear plant; then it spread over all Northern Europe (Scandinavia, Netherlands, Belgium and England), reaching down North-Eastern-Italy and the Mediterranean Sea ten days after the accident.

Emergency squads of workers and firefighters worked for several days to contain the fire and the spread of radioactive substances, in particular by dumping from helicopters mixtures of sand, clay, lead and boron (which is a neutron-absorbing material) over the fire, eventually stopping both the fire and the release of radioactive isotopes. The destroyed unit 4 was enclosed in a concrete shelter, which is being replaced, with the help on an international cooperation, by a more permanent structure to be completed in 2017.

During the Chernobyl nuclear plant disaster, people were evacuated (many permanently) from a zone of roughly 30 km. Given the major release of radioactive material and the widespread effects, the accident was rated at 7 on the INES.

The Chernobyl disaster was the result of major design deficiencies in the RBMK type of reactors (in particular the absence of the containment structure), the violation of operating procedures and the insufficient safety culture. Investigations following the accident underlined the vital importance of the defence-in-depth concept. Based on the analysis of the accident, the existing reactors of this type underwent changes and safety upgrades. While the RBMKs in Ukraine and

Lithuania were permanently shut down in 2000 and 2009, respectively, 11 units are still active in the Russian Federation.

The isotopic releases from the reactor included about 60 radionuclides. According to a study of the NEA, the total release, measured in ^{131}I equivalent release activity, amounted to 5.16×10^{18} Bq, a value actually more than 150 times higher than the INES Level 7 criterion.

The accident caused the death, within a few days or weeks, of 31 power plant employees and firefighters (including 28 with acute radiation syndrome). It also caused radiation sickness in a further 237 staff and firefighters (acute radiation syndrome diagnoses were later confirmed for 134 people) and brought about the evacuation, in 1986, of about 116,000 people from areas surrounding the reactor and the relocation, after 1986, of about 220,000 people from Belarus, Russia and Ukraine. Vast territories of those three countries were contaminated, and trace deposition of released radionuclides was measurable in all countries of the northern hemisphere. In addition, 19 highly-exposed persons died between 1987 and 2004 but the connection to radiation exposure is not certain [15].

About 130,000 people received significant radiation doses (i.e. above internationally accepted limits established by the International Commission on Radiation Protection—ICRP). Large scale investigations on the corresponding health effects were conducted in particular by the World Health Organization for at least twenty years [16]. In the population at large, no cases of acute radiation effects were reported. However, about 4000 cases of thyroid cancers in children were diagnosed since the accident, which can be reasonably connected to the intake of radioactive iodine released by the destroyed reactor. No increase of leukaemia or other types of cancer have yet shown up, but some is expected based on conservative probabilistic assumptions.

5.9.6   Vandellòs (1989), Spain

The accident at the gas-cooled, graphite-moderated reactor of Vandellòs in Catalonia, Spain, occurred on October 19, 1989 when a fire started in the generator due to a mechanical failure. The fire continued for almost four and a half hours and seriously affected two of the four connections from the turbine to the plant's cooling system. The fire impaired important nuclear safety functions in the plant.

The incident was classified as level 3 on the INES. The damage suffered by the safety systems caused a degradation of the defence-in-depth at the plant, but there was no leakage of radioactive materials to the outside.

5.9.7   Tokai-Mura (1999), Japan

The first serious nuclear accident in Japan's history occurred in 1999 in a conversion test building of a uranium reprocessing facility located in Tokai-Mura, approximately 120 km northeast of Tokyo. The accident was caused by bringing together too much uranium enriched to a relatively high level, causing a limited uncontrolled nuclear chain reaction, which continued intermittently for 20 h. Two workers died and 119 people were exposed to high levels of radiation.

The accident was rated 4 in the INES. According to the IAEA the cause of the accident was "human error and serious breaches of safety principles".

5.9.8   Davis-Besse (2002), USA

The Davis-Besse Nuclear Power Station, situated northeast of Oak Harbor in Ottawa County, Ohio, USA, has a single pressurised water reactor. On March 5, 2002, it was discovered that the borated water that serves as the reactor coolant had leaked from cracked control rod drive mechanisms directly above the reactor and eaten through more than 150 mm of the carbon steel reactor pressure vessel head over an area roughly the size of a football. Although the corrosion did not lead to an accident, this was considered a serious nuclear safety incident classified at level 3 on the INES.

5.9.9   Paks (2003), Hungary

On April 10, 2003, severe damage to fuel assemblies took place during an incident at Unit 2 of the nuclear power plant located near Paks, at 100 km south of Budapest, in Hungary. A majority of 30 fuel assemblies, which were being cleaned in a special tank below the water level of the spent fuel storage pool in order to remove crud buildup, got severely damaged and radioactivity was released into the cleaning tank. The incident was classified at level 3 of the INES.

5.9.10   Fukushima Daiichi (2011), Japan

The reported information about the accident is based on publications by NEA [12] and WNA [17].

The Fukushima Daiichi Nuclear Power Plant, located in the Futaba District of Fukushima Prefecture, on the Honshu Island (the main part of Japan), comprised six

BWRs, for a total power of 4.7 GW(e) and was first commissioned in 1971. At the time of the accident, the city had an estimated population of 290,064 people.

On March 11, 2011, a great earthquake (magnitude 9 on the Richter's scale, the largest ever recorded in Japan, later named the "Great East Japan Earthquake") and a subsequent gigantic anomalous wave, a so-called tsunami, hit the Tohoku region where Fukushima is located.[5] At the moment of the earthquake, only units 1–3 were operating and they shut down automatically and safely. Units 4, 5 and 6 were not operative because of periodic inspections. The fuel from unit 4 had been transferred to the spent fuel pool, while units 5 and 6 were in cold shutdown (the term is used to define a reactor coolant system at atmospheric pressure and at a temperature below 100 °C following a reactor cool down).

Because of the earthquake, there was a blackout of the outside power grid so that external power to the plant was interrupted. As foreseen in the automatic procedures, the emergency diesel generators switched on and started to provide the backup electric power needed to keep removing the residual decay heat from the core via the emergency cooling systems. But about 1 h after the earthquake, a tsunami, later reported to have been more than 14 m high, hit the site, heavily flooding all the area of the plant. The flooding was so extensive that all the diesel generators in the plant except one (enough only to protect the units 5 and 6 in cold shutdown) were covered by the sea water and stopped working, along with the pumps providing cool water from the ocean (this is the last level of thermal energy removal from the plant, also called the ultimate heat sink). Backup batteries also stopped supplying electric power after their lifetime of a few hours. Hence all safety systems guaranteeing the first two levels of defence-in-depth (fuel matrix and fuel cladding integrity) in units 1, 2 and 3 started to be compromised. Therefore, even though the earthquake itself, in spite of being so powerful, had apparently not caused damage to the reactors, the impact of the tsunami swept away one of the fundamental safety functions, namely decay heat removal from the core and its release to the ultimate heat sink.

After some time from when the safety systems stopped working, the reactor cores started to be compromised. Subsequent estimates put core melting in unit 1 at

---

[5]The earthquake was a rare and complex double quake that lasted about 3 min. An area of the seafloor extending 650 km north-south moved typically 10–20 m horizontally. Japan moved a few metres east and the local coastline subsided half a metre. The tsunami inundated an area of about 560 km^2 and resulted in a human death toll of over 15,892, while 2573 persons were missing and 6135 were injured as of August 10, 2015, according to Japan's Reconstruction Agency [18]. Much damage was done to coastal ports and towns with over a million buildings destroyed or partly collapsed. Most people died by drowning. In total, over 470,000 people were evacuated from their homes because of earthquake, tsunami and nuclear accident (about one third of the evacuations were related to the latter). As of August 2015, the number of evacuees had decreased to less than 200,000 people, among which 70,000 are still in temporary housings.

The earthquake shifted the Earth on its axis of rotation by redistributing mass, like putting a dent in a spinning top. It also shortened the length of a day by about a microsecond. The effects of the earthquake were felt around the world, from Norway's fjords to Antarctica's ice sheet. Tsunami debris continued to wash up on North American beaches two years later [19].

several hours after the arrival of the tsunami, in unit 2 on the following day, and in unit 3 about 2 days later. Together with the core melting, the chemical reaction of the water steam with the zirconium in the cladding at very high temperatures produced a significant quantity of hydrogen gas. Evacuations were ordered by the authorities as soon as the potential consequences of the accident became evident. As soon as a day after the accident had started, an evacuation zone of 20 km from the plant was established. In the area between 20 and 30 km from the plant, evacuation was ordered when the dose rate exceeded 20 mSv/a. Such limit is used now to determine whether return of the population to their homes can be allowed.

In order to relieve the high pressure within the reactor vessels and avoid damage to the primary containments, the ventilation system was used to vent the gases outside of the containment. However, some of the hydrogen gas accumulated within the reactor buildings (secondary containment) of all four units 1–4, and eventually ignited producing explosions that severely damaged the reactor buildings. Because of the extensive damage to the fuel elements, large amounts of radioactive isotopes with different degree of volatility were free to move from the reactor core. Ventilation to relieve the pressure would have vented outside also such radioactive elements, but in a controlled way. Instead, the explosions led to an uncontrolled release of large quantities of radioactive nuclides into the outside environment.

Considering two of the most radiologically significant isotopes, Cs-137 and I-131, the corresponding releases in the first 20 days after the accident (i.e. until the end of March) were estimated in 2012 by the company operating the plant at 10^{16} Bq and 5×10^{17} Bq, respectively, about 12 and 28 %, respectively, of the amount spread out in the Chernobyl accident. Further releases that occurred later, from April to the end of 2011, were estimated to be less than 1 % of the above amounts. Given that these values are about 30 times the INES Level 7 criterion, the accident at Fukushima Daiichi received an overall rating of 7 on that scale, the same as Chernobyl.

At the time of writing this book, thanks to the prompt evacuation procedures, no casualties or acute radiation syndromes have been observed as a consequence of the accident. However, some workers were exposed to high radiation levels. By assuming that the probability to develop a disease has a linear relationship with the absorbed dose (see Sect. 2.10), it has been estimated that a limited number of cancer cases may develop within the local exposed population.[6] On the other hand,

---

[6]The 2013 report [19] from the World Health Organization states that "The results show the largest additional cancer risks among those exposed in infancy (leukaemia in males and solid cancers in females). Given the exposure to radioactive iodine, during the early phase of the emergency, the lifetime attributable risk of thyroid cancer was specifically assessed. The results show the greatest risk among girls exposed as infants in the most affected area in Fukushima prefecture, although the excess absolute risk is small, because of the low baseline risk of thyroid cancer, it represents a comparatively high relative increase in the lifetime risk of up to around 70 % (as an upper bound). The high relative risk of childhood thyroid cancer becomes more evident when risks are calculated over the first 15 years after the accident for those exposed as infants, because the baseline thyroid cancer risk in early life is very low. Monitoring children's health is therefore warranted. The risk of

as in all similar cases, it will be very difficult to associate any modest increase in the number of cancer cases to the radiation exposure, except for some types of diseases more directly related to the exposure, like for instance thyroid nodules and cancers in children. It is also important to remark that, as a consequence of the Fukushima Daiichi accident, a certain quantity of food and fish in the area were contaminated with a significant impact on the local economy. Such contamination appears greatly reduced as of today: according to dedicated research, during the first year after the accident, 3.3 % of food from the Fukushima region had above-limit contamination (these products were prevented by the authorities from reaching the market). The percentage rose slightly the following year but by 2014, the proportion had fallen to 0.6 %. For the whole Japan, the figures were 0.9 % falling to 0.2 % [20].

Work at international level to analyse the data collected, fully understand the accident and its consequences and draw lessons from it is still ongoing and will likely continue for a while. In September 2015 the IAEA published a comprehensive report along with 5 technical volumes on the causes and consequences of the accident [21]. The report is the result of an extensive collaboration that involved some 180 experts from 42 IAEA Member States and several international bodies. The report considers human, organizational and technical factors, and aims to provide an understanding of what happened, and why, so that the necessary lessons learned can be acted upon by governments, regulators and nuclear power plant operators throughout the world. Measures taken in response to the accident, both in Japan and internationally are also examined.

Among the issues identified by the IAEA and governmental reports on the event are organisational deficiencies, bureaucracy, unclear definitions of responsibility between the operator, the regulator and the government, and failure to properly implement lessons learnt from the TMI and Chernobyl accidents.

All over the world, this accident has prompted:

- a halt in the operation of several plants. In Japan itself, all plants were shut down for inspections and re-examination of the licenses and at the time of writing this book, only two reactors have restarted operation. In the whole European Union, a series of stress tests have been conducted to double check the capability of the

---

(Footnote 6 continued)

leukaemia as a result of radiation exposure from the accident was assessed to be greatest in males exposed as infants in geographical locations with the highest exposure, slightly above 5 % over baseline risk as an upper bound. A similar result is found for breast cancer in girls exposed as infants. For all solid cancers, a maximum relative increase of about 4 % was estimated." About workers at the plant, the report states that "To date, the Fukushima Daiichi NPP accident has not resulted in acute radiation effects among workers. None of the seven reported deaths among workers is attributable to radiation exposure. For about one third of the workers, the relative increase over background for thyroid cancer is estimated to be up to 20 % for the youngest workers. For less than 1 % of workers, the relative increase over background for leukaemia and thyroid cancer is as high as 28 % in the youngest workers. For those few emergency workers who received very high doses to the thyroid, a notable risk of thyroid cancer is estimated, especially for young workers.".

plants to withstand extreme events from earthquakes to very severe weather and flooding; in Germany, a few reactors have been permanently shut down to undergo decommissioning and dismantling in view of abandoning nuclear power in a few years from now.

- The decision of a few countries to establish a schedule for abandoning nuclear energy production (Belgium, Germany), to stop construction of new reactors (Switzerland) or to cancel a previous governmental decision to restart nuclear power (Italy).
- The establishment of a number of review committees for the revision of operation, safety procedures, design basis accidents and design of new reactors; it is worth mentioning the IAEA Action Plan on Nuclear Safety of 2011 [22].
- A strong political and social debate practically everywhere in the world, with a strong pressure from the public to reconsider the option of nuclear power.

Analyses of the accident showed the need for more careful siting criteria than those used in the 1960s (that is at the time of the construction of the plant[7]). They also showed that a better back-up power and post shutdown cooling was necessary, and that better provision had to be made for venting the containment of that kind of reactor as well as for other emergency management procedures.

Problem 5.1: Iodine and Caesium Contamination. In a nuclear accident the principal health hazard is from the spread of radioactive materials, notably volatile fission products such as ^{131}I (half-life about 8 days) and ^{137}Cs (half-life about 30 years). The given radioactive products are biological active, so that if consumed in food, they tend to accumulate in the organs of the body. Radioactive iodine enters the body and accumulates in the thyroid gland, and apparently gave rise to the thyroid cancers after the Chernobyl accident. Cs-137 has a half-life of 30 years, and is therefore a potential long-term contaminant of pasture and crops. Which measures can be taken to limit their hazard (see Problem 2.9)?

Solution: Due to the short life-time, ^{131}I is a hazard for around the first two months when its radioactivity reduces by a factor 256. Then, to limit human uptake of ^{131}I one can evacuate the contaminated area for several weeks, and assume potassium iodide tablets containing stable iodine to saturate the thyroid gland so that it cannot absorb any more iodine—either stable or radioactive. Instead, due to its long half-life, high levels of radioactive caesium can preclude food production from affected land for a long time. In this case, for instance, a measure to adopt would be to check the milk (with particular attention to children as the most affected category in the population) and in case of radioactivity above the limits, stop its production and sale.

---

[7]The tsunami heights coming ashore were about 14 m high, with respect to the design basis value of 5.7 m.

Problem 5.2: Chernobyl Accident. In the Chernobyl accident, reactor output suddenly increased from about $P_1 = 0.20 P_0$ to $P_2 = 100 P_0$, where $P_0 = 3$ GW is the nominal power of the reactor, in about $\Delta t = 4$ s. Assuming a constant reactivity excess during this time, calculate what was the average reactor period T_R and how much energy was released during this time.
Solution: The power of the reactor rises as the number of neutrons, that is, using Eq. (3.21), as

$$P_2 = P_1 e^{\Delta t / T_R}.$$

From which it is easy to obtain

$$T_R = \frac{\Delta t}{\ln \frac{P_2}{P_1}} = \frac{4}{\ln \frac{100}{0.20}} = 0.64 \, \text{s}.$$

The energy released in the time Δt is

$$E = P_1 \int_0^{\Delta t} e^{t/T_R} dt = P_1 T_R (e^{\Delta t / T_R} - 1) = 0.2 \times 3 \times 10^9 \times 0.64 \times (e^{4/0.64} - 1) = 198.5 \, \text{GJ}.$$

5.10   Safety Relative to Other Energy Sources

The few and rare accidents occurred at commercial nuclear power plants have shown that these plants are complex systems vulnerable to accidents and failures because of natural disasters such as flooding, earthquakes and extreme weather, and because of fires, equipment failures, improper maintenance, and human errors. However, in considering various energy sources, it is necessary to properly put the risks in perspective, keeping in mind that any form of energy production, and in general any human activity, has its drawbacks and risks, and that severe accidents causing a large number of fatalities have also occurred in other major energy production chains.

In [23], OECD presented an analysis compiled by the Paul Scherrer Institute on every accident causing five or more deaths in the energy industry between 1969 and 2000. Table 5.3 summarises the results: in the examined time period there were in total 1870 such severe accidents worldwide resulting in 81,258 immediate deaths. The worst energy-related accident was the Banqiao/Scimantan dam failure in China in 1975 when some 30,000 people were killed immediately and some 230,000 overall. Among the fossil chains, coal accounts for most deaths, followed by oil, liquefied petroleum gas (LPG) and natural gas. At the time of the study, the only

Table 5.3 Summary of severe (≥ 5 deaths) accidents that occurred in fossil, hydro and nuclear energy sectors in the period 1969–2000 [23]

Energy chain	OECD countries			Non-OECD countries		
	Accidents	Deaths	Deaths/TWa	Accidents	Deaths	Deaths/TWa
Coal	75	2259	157	1044	18,017	597
Coal (data for China 1994–1999)				819	11,334	6169
Coal (without China)				102	4831	597
Oil	165	3713	132	232	16,505	897
Natural gas	90	1043	85	45	1000	111
LPG	59	1905	1957	46	2016	14,896
Hydropower	1	14	3	10	29,924	10,285
Nuclear	0	0	0	1	31	48
Total	390	8934		1480	72,324	

Deaths are normalised to the produced energy (deaths per terawatt-year [TWa])

severe accident at a nuclear plant was the Chernobyl accident that killed 31 plant and emergency workers, shortly after the event.

OECD countries exhibit significantly lower death rates per unit of energy generated than non-OECD countries for all energy chains. In the fossil chains, LPG has the highest death rate (when dividing the number of deaths by the amount of produced energy in TW-years, TWa), followed by coal, oil and natural gas. OECD nuclear and hydropower plants have the lowest fatality rates, while in non-OECD countries historical evidence suggests that dam failures pose a much higher risk. Table 5.3 also shows that Chinese data for coal should be treated separately as its accident death rates are about ten times higher than in other non-OECD countries and about forty times higher than in OECD countries [23, 24].

For the nuclear accident at Chernobyl immediate fatalities are less significant than so-called *latent fatalities*, i.e. delayed fatalities arising many years later caused by exposure to released radioactive material. Clearly, the latter are more difficult to estimate. Studies by the European Community, the IAEA, the WO and UNSCEAR estimated latent deaths following the Chernobyl accident to be between 9000 (based on dose cut-off) and 33,000 (entire northern hemisphere with no cut-off) over the next 70 years. This is equivalent to between 1390 and 5120 deaths per terawatt-year for non-OECD countries. However, extrapolating these nuclear energy risks to the current fleet of nuclear reactors is not appropriate, because of the overall evolution towards safer technologies and stricter regulatory frameworks than were in force in the Ukraine at the time of the Chernobyl accident.

For a correct comparison one should also consider the latent health effects of fossil fuel burning, the main alternative for baseload electricity production. An OECD study in 2008 [25] reports that outdoor air pollution due to fine particulate matter ($\leq 10\,\mu$) is estimated to have caused approximately 960,000 premature deaths in the year 2000 alone and 9.6 million years of life lost world-wide. Of this pollution, about 30 % arises from energy sources. Hence, even on a latent deaths basis, the outcome of the Chernobyl accident is small in comparison with those of other energy sources, predominantly fossil fuel burning. Overall, accident related deaths per unit energy production are much smaller for nuclear power than those resulting from the health effects of fossil fuel emissions, but they attract much more media and public attention.

Nuclear accidents have occurred with a much lower probability than in other industrial sectors but, at the same time, a single nuclear accident can have an enormous impact on people's life and wellbeing. Because of this specific aspect, the choice on whether to use nuclear fission energy or not has become more of a social and ethical issue than a scientific one.

This is why safety measures and a safety culture have an immense importance in commercial energy production from the nuclear source. Indeed, the philosophy underlying upgrades in Generation III reactors, new projects in Generation III+ and also the Generation IV, that may replace the current ones in the future, is to continue to improve on the reliability and soundness of the safety design (see Sect. 4.5). Particular emphasis is put on aspects like redundancy, independence and diversity, as well on the adoption of passive systems, such as to guarantee and reinforce the defence-in-depth, reduce the probability of severe accidents even further and protect the people, the environment and the plants themselves.

5.11 Nuclear Security and Safeguards

Another major concern about the use of nuclear fission is that the expansion of nuclear technology may lead to the increased risk of misuse. Nuclear security deals with the prevention and detection of, and response to: theft, sabotage, unauthorized access, illegal transfer or other malicious acts involving nuclear material, other radioactive substances or their associated facilities or activities. Security focuses on the intentional misuse of nuclear and other radioactive materials. It relates mainly to external threats to materials or facilities. Another important aspect is to prevent nuclear weapons materials and technology to spread and become available to more entities.

Nuclear security is so far a national responsibility, which makes more difficult the implementation and assessment of standard international practices. In recent years, with the increasing terrorism threat, nuclear security has acquired more and more importance. This led the international community to make greater efforts to minimize this threat. As a result, treaties have been signed to this effect, among which it is worth to mention the Treaty on the Non-proliferation of Nuclear Weapons (NPT) [26] and the Comprehensive Nuclear Test Ban Treaty (CTBT) [27]. The intent of these treaties is to provide a basis of international agreements to prevent nuclear weapons from further spreading worldwide.

The NPT entered into force in 1970 and was extended indefinitely in 1995. In the document, a distinction is made between countries that produced and exploded nuclear devices before January 1, 1967 (namely China, France, the Russian Federation, the United Kingdom and the United States, the treaty calls them "nuclear weapon States"), and all the others. What the treaty asks of the nuclear weapons states is that they do not provide nuclear weapons to any of the other countries, nor assist them in developing nuclear weapons. What it asks of all the others is a commitment not to develop nuclear weapons and to use nuclear technology only for peaceful purposes. In addition, the treaty requires all countries to pursue nuclear disarmament. Almost all world countries signed the NPT, except India, Israel and Pakistan, while North Korea signed it, but withdrew from the treaty in 2003.

The IAEA has an essential verification role under the NPT, despite not being a party to the Treaty. In particular, under Article III of the Treaty, each Non-Nuclear-Weapon State Party is required to conclude a comprehensive "safeguards agreement" with the IAEA to enable it to verify the fulfilment of the State obligation to not divert nuclear material from peaceful activities to nuclear weapons manufacturing. Safeguards foresee that a country must declare the inventory and location of weapons-usable nuclear materials and where it is located. They also foresee the verification of a nuclear installation's control of and accounting for weapons-usable nuclear materials within all the nuclear facilities that a signatory State has formally declared as subject to safeguards. Verification is performed using IAEA-installed monitoring instruments, some of which are sealed to prevent tampering. Physical inspection of nuclear installations on a random, yet pre-announced, basis is conducted regularly to verify the operator's accounts and to ensure that all installed instruments are performing satisfactorily and that security seals have not been tampered with. Since 1997, IAEA inspections can also be carried out on a sudden or challenge basis once a State has ratified an additional safeguards protocol. The intended result of all inspections is that, by verifying the inventories of nuclear material declared by a signatory government, the IAEA can announce that all nuclear material is being used for peaceful purposes.

182 States have safeguards agreements in force with the IAEA. The vast majority of these agreements are those that have been concluded by the IAEA with non-nuclear-weapon States that are parties to the NPT. In 2014 there were more than 1250 nuclear facilities and locations under safeguards, and some 193,500 significant quantities of nuclear material under safeguards (a significant quantity is the approximate amount of nuclear material for which the possibility of manufacturing a nuclear explosive device cannot be excluded).

The Comprehensive Nuclear Test Ban Treaty (CTBT) was adopted by the United Nations General Assembly in 1996 but has not entered into force yet. Its essential goal is to forbid all nuclear explosions, either of military or civilian nature. The document foresees the possibility to perform on-site inspections, contains provisions for consultation and clarification, as well as measures to increase reciprocal confidence, all part of a general strategy for verification of the fulfilment of the obligations therein. As of March 2015, 164 countries have ratified the CTBT, while another 19 have signed it but not ratified it. However, the treaty will only become effective when it will be ratified by all 44 states that participated in the discussions to elaborate the CTBT between 1994 and 1996 and that were hosting nuclear power reactors or research reactors at that time. As of 2015, only 36 countries out of these 44 had ratified the treaty. China, Egypt, Iran, Israel and the United States have signed but not ratified the CTBT, while India, North Korea and Pakistan have not signed it.

The fear of the possible proliferation of nuclear weapons has been a major deterrent that has contributed to slowing down the expansion of nuclear energy in the past decades. For the same reason, already by the 1970s, the USA were opposed to fuel reprocessing as a possible source of diversion of uranium and plutonium for military use.

One of the major difficulties in producing an atomic bomb is in obtaining fissionable material, which may be U-235 or Pu-239, in suitable quantity and purity. In practice, one distinguishes between reactor-grade fuel, used in power reactors for electricity generation, and weapons-grade material. While reactor-grade uranium is typically enriched to about 3–4 % in U-235, weapons-grade uranium is enriched to over 90 %. The technology to produce weapons-grade uranium or plutonium is complex and expensive, so that it is not within reach of every country. However, it is not impossible to acquire it, given sufficient time and effort. Therefore, international cooperation and surveillance (the above mentioned safeguards regime) is crucial in maintaining control over fissile material resources and their final use.

Plutonium-239 is produced in varying quantities in virtually all operating nuclear reactors (especially breeder reactors) when uranium-238 absorbs a neutron and then quickly decays to neptunium, and then to plutonium. However, this breeding process is complex, because Pu-239 cannot stay for long time in the reactor, otherwise it will absorb additional neutrons, gradually transforming into the isotopes Pu-240, Pu-241, and so on. These other isotopes present a high spontaneous fission rate and high gamma radiation levels, making plutonium from civil spent fuel difficult to use for weapons. In fact, the high spontaneous fission rate makes a weapon less stable, while the high gamma rate makes such plutonium mix more dangerous to handle. Therefore, in principle, only plutonium containing more than 93 % of the isotope Pu-239 is considered truly weapons-grade, and its production needs dedicated reactors with frequent fuel replacement.

At the contrary, the plutonium from spent fuel from regular discharges of a thermal reactor typically contains about 60–70 % Pu-239. The latter can be recycled as nuclear fuel by performing spent fuel reprocessing to separate all plutonium isotopes from the fission products and the heavier nuclides americium, curium, etc. (see Sect. 3.12). However, in spite of the above simple considerations, the assessment of the proliferation risk within a specific fuel cycle is a complex process, which depends on many factors, like production techniques, isotopic and chemical composition of fresh and spent fuel, type of the facilities used, etc. Therefore, the IAEA conservatively emphasizes the need to consider weapons proliferation risks on a case-by-case basis, with a detailed analysis of all the above aspects of the fuel cycle [28].

Problem 5.3: Hiroshima Bomb. The explosive yield of TNT (trinitro-toluene) is considered the standard measure for strength of bombs and other explosives, with 1 tonne of TNT equalling 4.184×10^9 J. The nuclear bomb that the USA dropped over Hiroshima on August 6, 1945 during the final stage of World War II was about 15 kt of TNT. Knowing that the bomb contained $M = 64$ kg of ^{235}U, calculate what fraction of the uranium actually underwent fission. Assume an average energy release per fission of 200 MeV.
Solution: The energy released by the explosion was $E = 15 \times 10^3 \times 4.184 \times 10^9 = 6.28 \times 10^{13}$ J. Then, calling $E_0 = 200$ MeV $= 3.2 \times 10^{-11}$ J the energy released in a single fission, $N_A = 6.022 \times 10^{23}$ the Avogadro number and $M_U = 0.235$ kg/mol the molar mass of ^{235}U, the mass m of ^{235}U that has undergone fission is

$$ m = \frac{E}{E_0}\frac{M_U}{N_A} = \frac{6.28 \times 10^{13}}{3.2 \times 10^{-11}}\frac{0.235}{6.022 \times 10^{23}} = 0.766\,\text{kg}. $$

Then only about $0.766/64 = 1.2$ % of the uranium in the bomb underwent fission.

Problem 5.4: Natural radioactivity. An isolated rock of mass 100 tonnes contains 0.1 % in weight of ^{238}U. As shown in Fig. 2.2, ^{238}U is the parent of a decay series (half-life 4.47×10^9 a) developing ^{226}Ra which, in turn, undergoes α decays (half-life 1.60×10^3 a) producing ^{4}He. Calculate the amount of radon in the rock and the rate of helium production in grams per year.
Solution: the radium is much shorter-lived than uranium ($\tau_{1/2\text{Ra}} \ll \tau_{1/2\text{U}}$) and clearly a long time has passed since the formation of the rock, then from Eq. (2.23) it follows the secular equilibrium condition

$$ \frac{N_{Ra}}{N_U} = \frac{\tau_{1/2\text{Ra}}}{\tau_{1/2\text{U}} - \tau_{1/2\text{Ra}}} = \left(\text{being } \tau_{1/2\text{Ra}} < <\tau_{1/2\text{U}}\right) = \frac{\tau_{1/2\text{Ra}}}{\tau_{1/2\text{U}}}, $$

where N_{Ra} and N_U are the number of nuclei of radon and uranium, respectively. Calling M_{Ra} and M_U the masses of radon and uranium, respectively, and $P_{Ra} = 226$ and $P_U = 238$ their atomic masses, the above ratio is also given by

$$ \frac{N_{Ra}}{N_U} = \frac{M_{Ra}}{M_U}\frac{P_U}{P_{Ra}}, $$

Combining the above two equations one has

$$ M_{Ra} = M_U \frac{P_{Ra}}{P_U}\frac{\tau_{1/2\text{Ra}}}{\tau_{1/2\text{U}}} = 10^5 \times 0.1 \times 10^{-2}\frac{226}{238}\frac{1600}{4.47 \times 10^9} = 3.4 \times 10^{-5}\text{kg}. $$

The rate of helium production is equal to the activity of radon (see Eq. (2.12))

$$\frac{dN_{He}}{dt} = \lambda_{Ra} N_{Ra}$$

Which, written in terms of the masses M_{Ra} and M_{He} of radon and helium nuclei (calling $P_{He} = 4$ the atomic mass of ^{4}He), becomes

$$\frac{dM_{He}}{dt} = \lambda_{Ra} M_{Ra} \frac{P_{He}}{P_{Ra}} = M_{Ra} \frac{0.693}{\tau_{1/2Ra}} \frac{P_{He}}{P_{Ra}} = 3.4 \times 10^{-5} \frac{0.693}{1600} \frac{4}{226}$$
$$= 2.6 \times 10^{-10} \, \text{kg/a}.$$

Problem 5.5: Radon Accumulation. A hotel meeting room, sized $10 \times 10 \times 4$ m^3 and with walls, floor and ceiling made of concrete, is not aerated for several days after a convention. The measured specific activity of ^{222}Rn is 100 Bq/m^3. Determine the concentration of ^{238}U in the concrete in nuclei/m^3 and in gr/m^3, if the effective thickness from which the radon, which originates from the α decay of uranium, can spread in the room is equal to 1.5 cm. Assume the migration of the radon from the concrete to the room to be complete once the measurement is performed.
Solution: The volume of the room is $V = 10 \times 10 \times 4 = 400$ m^3, then the activity of radon is

$$R_{Rn} = 100 \, V = 4 \times 10^4 \, \text{Bq}.$$

The radon-222 (half-life about 3.82 d) is much shorter-lived than uranium-238 (4.47×10^9 a), then we can apply Eq. (2.26) which says that the activity of the two nuclides are equal.

$$N_{Rn} \lambda_{Rn} = N_U \lambda_U.$$

where N_{Rn} and N_U are the number of nuclei of Rn and U, respectively. From this it follows

$$N_U = N_{Ra} \frac{\lambda_{Ra}}{\lambda_U} = \frac{R_{Ra}}{\lambda_U} = R_{Ra} \frac{\tau_{1/2U}}{0.693} = 4 \times 10^4 \times \frac{4.47 \times 10^9 \times 3.15 \times 10^7}{0.693}$$
$$= 8.13 \times 10^{21} \, \text{nuclei}.$$

The volume of concrete w from which radon originates is equal to the total area of walls, floor and ceiling multiplied by the thickness of 15 cm

$$w = 4 \times (10 \times 4) \times 0.015 + 2 \times 10 \times 10 \times 0.015 = (160 + 200) \times 0.015$$
$$= 5.40 \, \text{m}^3.$$

Then the concentration of U nuclei in the concrete is

$$\frac{N_U}{w} = \frac{8.13 \times 10^{21}}{5.4} = 1.5 \times 10^{21} \text{nuclei}/\text{m}^3,$$

$$\frac{m}{V} = \frac{1.5 \times 10^{21}}{6.022 \times 10^{23}} \times 238 = 0.59 \,\text{gr}/\text{m}^3.$$

Problems

5.1. The radioactive isotope strontium-90 (half-life of 28.8 years) is one of the waste products produced in nuclear reactors. Calculate how long would one have to wait for the radioactivity from a sample of Sr-90 to become (a) 1000 times smaller, (b) one million times smaller.
[*Ans.*: (a) 287 a; (b) 574 a]

5.2. In a nuclear accident the principal health hazard is from the spread of radioactive materials, notably volatile fission products such as ^{131}I (half-life about 8 days) and the two caesium isotopes ^{134}Cs (half-life about 2 years) and ^{137}Cs (half-life about 30 years). Calculate how long would one have to wait for the radioactivity from these nuclides to become (a) 1000 times smaller, (b) one million times smaller.
[*Ans.*: (a) 80 d for ^{131}I, 20 a for ^{134}Cs, and 299 a for ^{137}Cs; (b) 159 d for ^{131}I, 40 a for ^{134}Cs and 598 a for ^{137}Cs]

References

1. https://www.iaea.org/publications/documents/treaties/convention-nuclear-safety
2. IAEA document Safety Fundamentals, The safety of nuclear installations, (IAEA Safety Series No. 110 published 1993)
3. IAEA Safety Glossary, Terminology used in nuclear safety and radiation protection, 2007 Edition
4. Basic Safety Principles for Nuclear Power Plants (INSAG-3), http://www-pub.iaea.org/books/IAEABooks/5811/Basic-Safety-Principles-for-Nuclear-Power-Plants-75-INSAG-3-Rev-1
5. NEA-6885 Nuclear energy today, ISBN 978-92-64-99204-7, OECD 2012
6. World Nuclear Association, Source safety of nuclear power reactors. August 2015, http://world-nuclear.org/info/Safety-and-Security/Safety-of-Plants/Safety-of-Nuclear-Power-Reactors
7. IAEA Power Reactor Information System, http://www.iaea.org/pris
8. Nuclear Power Plant operating experience, from the IAEA/NEA international reporting system for operating experience, 2009–2011, http://www-ns.iaea.org/downloads/ni/irs/npp-op-ex-2009-2011.pdf
9. http://www.wano.info/en-gb
10. http://www-pub.iaea.org/MTCD/Publications/PDF/INES2013web.pdf
11. Z.A. Medvedev, Two Decades of Dissidence, New Scientist, 4 Nov 1976
12. Nuclear energy today, ISBN 978-92-64-99204-7, © OECD 2012

13. http://www.world-nuclear.org/info/Safety-and-Security/Safety-of-Plants/Three-Mile-Island-accident
14. http://www.world-nuclear.org/info/Safety-and-Security/Safety-of-Plants/Chernobyl-Accident
15. http://www.world-nuclear.org/information-library/safety-and-security/safety-of-plants/appendices/chernobyl-accident-appendix-2-health-impacts.aspx
16. World Health Organization Report, http://apps.who.int/iris/bitstream/10665/43447/1/9241594179_eng.pdf
17. http://www.world-nuclear.org/info/Safety-and-Security/Safety-of-Plants/Fukushima-Accident
18. http://www.reconstruction.go.jp/english/topics/GEJE/index.html
19. Health risk assessment from the nuclear accident after the 2011 Great East Japan earthquake and tsunami, © World Health Organization 2013. http://apps.who.int/iris/bitstream/10665/78218/1/9789241505130_eng.pdf
20. E. Gibney, Fukushima data show rise and fall in food radioactivity, Nature news, 27 February 2015, http://www.nature.com/news/fukushima-data-show-rise-and-fall-in-food-radioactivity-1.17016
21. The Fukushima Daiichi Accident: report by the director general and five technical volumes, STI/PUB/1710 (ISBN: 978-92-0-107015-9). http://www-pub.iaea.org/books/IAEABooks/10962/The-Fukushima-Daiichi-Accident
22. https://www.iaea.org/newscenter/focus/nuclear-safety-action-plan]
23. Comparing nuclear accident risks with those from other energy sources, OECD 2010, NEA No. 6861
24. P. Burgherr, S. Hirschberg, Science direct, energy 33 (2008) 538. OECD-NEA 6862 comparing risks
25. OECD, OECD Environmental Outlook (OECD, Paris, 2008)
26. Treaty on the Non-Proliferation of Nuclear Weapons (NPT), United Nations Office for Disarmament Affairs
27. Resolution 50/245 adopted by the general assembly of United Nations at its 50th session. "Comprehensive nuclear-test-ban treaty". United Nations, 17 September 1996. Retrieved 3 Dec 2011. http://www.un.org/documents/ga/res/50/ares50-245.htm
28. Technical Features to Enhance Proliferation Resistance of Nuclear Energy Systems, IAEA nuclear energy series, IAEA, Vienna 2010, http://www-pub.iaea.org/MTCD/publications/PDF/Pub1464_web.pdf

Chapter 6
Management of Radioactive Waste

Electricity production from nuclear power does not emit greenhouse gases like carbon dioxide into the air. However, it generates radioactive wastes in solid, liquid, and gaseous forms that can contaminate the environment and create a hazard for people's health if not properly controlled and managed. Therefore, radioactive wastes must be handled in a safe way to protect people and the environment. Various types of radioactive waste exist and proper disposal depends on the properties of the waste. This chapter focuses on nuclear waste generation, handling, storage and disposal. It first gives the classification of nuclear wastes according to their radioactive content and half-life, making comparisons with other hazardous waste and waste from other sources of electricity generation. Then, it illustrates the typical composition of nuclear spent fuel, the time evolution of its radioactivity, and the safe methods for its final disposal. Finally, it describes the studies in progress for significantly reducing the volume and radiotoxicity of nuclear waste and for shortening the very long time for which they must be stored safely.

6.1 Types of Radioactive Waste

The term *radioactive waste* indicates all residual materials with some degree of radioactivity above the natural background, arising from any type of nuclear activity. The name waste implies that no further use of such materials is possible so that they have to be disposed of as it happens with other, non-radioactive waste. However, the radioactive nature of the nuclear waste requires special procedures in handling, conditioning, storing and disposing of the materials. In particular, in many cases special treatments and measures have to be adopted to isolate the material from the biosphere, as it happens for other types of toxic waste. Moreover, for radioactive waste one must consider the additional hazard stemming from the fact that the radiation can cause harm to the human body also from a distance, even

© Springer International Publishing Switzerland 2016
E. De Sanctis et al., *Energy from Nuclear Fission*, Undergraduate Lecture Notes in Physics, DOI 10.1007/978-3-319-30651-3_6

though the material is not touched, inhaled or ingested. The hazard is particularly high in case of nuclear wastes characterized by either a high level of radioactivity and/or the presence of long-lived radionuclides that make the radioactivity last for very long time.

Generally, in handling and storing radioactive waste, three methodologies are followed. First of all, techniques are employed whenever possible to reduce the volume of the wastes and put them in appropriate safe and secure containers. Secondly, when necessary and applicable, certain wastes resulting from industrial treatment processes are diluted and then discharged into the environment, by respecting strict safety limits on the quantity of radioactivity discharged. Finally, the only way the radioactive wastes become harmless is through radioactive decay; for high-level wastes, this can take hundreds or thousands of years. For such wastes, the international consensus is that geological repositories are required, the latter being suitable sites at a depth of several hundred metres below ground, offering specific safety features in terms of geological long term stability, interaction with underground water, etc.

Beyond electricity production, wastes containing radioactive substances are generated by applications of nuclear science and technology in fields of medicine, industry, scientific research, and the military. For instance, radioactive wastes are also generated: (*i*) in hospitals, to diagnose and treat diseases and to sterilize medical tools, (*ii*) in universities and research centres, to conduct research in physics, biology, chemistry and engineering, (*iii*) in agriculture, to produce crops that are more drought and disease resistant, as well as crops with shorter growing periods or increased yield—a practice that has been especially beneficial for some developing countries. Radioactive wastes also arise when nuclear facilities are decommissioned and permanently shut down.

Good waste management begins before the waste is generated: a fundamental principle for all activities that produce radioactive waste is to avoid or reduce waste generation at its source. Minimizing primary waste generation also minimizes the quantity of waste requiring storage and disposal.

Various types of radioactive waste exist and proper disposal will depend on the properties of the waste. Radioactive wastes are normally classified into a small number of categories to facilitate regulations about handling, storage and disposal, based on the radioactive content and thermal characteristics. The radioactive waste hazard determines the type of containment and isolation required.

Within the nuclear energy sector, a rough categorization divides nuclear waste into low-level wastes (usually abbreviated as LLW), intermediate-level wastes (ILW) and high-level wastes (HLW). A few countries have found it useful to introduce the category of very low-level waste (VLLW). This type of waste may be generated in relatively large volume, notably from the decommissioning of nuclear installations, but carries very low radiological hazards, to the point that it may be managed in facilities that do not require a nuclear license. This categorization varies slightly from country to country but, in principle, the main criteria for determining the type of waste are based on its radioactive content and the time needed for that radioactivity to decay to insignificant levels.

The IAEA also uses the following categories [1]: (a) exempt waste (EW) which is not relevant for radiation protection and can be released from regulatory controls; (b) very short-lived waste (VSLW) which can be stored for a few years so that the short-lived isotopes that it contains can decay enough to release the material from regulatory controls; (c) very low-level waste (VLLW), with a higher content of radioactive materials, but at a level which is not considered hazardous to people or the surrounding environment.

Low- and intermediate-level wastes arise mainly from routine facility maintenance and operations. For both these categories, the heat generated by the internal radioactivity must be below about 2 kW/m^3 [2]. Often a further distinction is made between:

(1) Short-lived waste (LILW-SL, i.e. Low and Intermediate Level Waste, Short-Lived), decaying in less than 30 years, where the content of all types of long-lived isotopes is limited to a maximum of 400 Bq/g [2],
(2) Long-lived waste (LILW-LL, i.e. Low and Intermediate Level Waste, Long-Lived), taking more than 30 years to decay, where the content of all types of long-lived isotopes exceeds 400 Bq/g [2].

Whatever the type of the radioactive waste, all of it has to be disposed of in a safe manner. Therefore, activities like handling or transporting the waste as well as designing storage or disposal sites have to be performed by applying fundamental safety principles, similarly to what done in designing and operating reactors. However, here the engineering aspects are less demanding, given the absence of reactor features like control of high power, complex cooling systems and intervention of feedback mechanisms.

6.1.1 Very Low-Level Waste (VLLW)

Very low-level wastes are wastes with a very limited content of long-lived isotopes and low levels of activity (activity of α emitters generally less than 10 Bq/g [2]), such that it does not need great care in isolating and containing it. VLLW can be disposed of in surface repositories of the landfill type and does not require extensive controls and surveillance. A common form of surface repository are trenches where the VLLW are placed.

A typical example of VLLW is demolished material with very low activity per unit mass, such as soil and rubble (concrete, plaster, bricks, metal, valves, pipes, etc.) produced during refurbishing or dismantling operations in nuclear sites.

6.1.2 Low-Level Waste (LLW)

Low-level wastes include items whose radioactivity stems from contamination (see the footnote in Sect. 2.10 for a definition of contamination). Examples of this type of waste are contaminated clothing, such as protective shoe covers, floor sweepings, cleaning cloths, paper towels, i.e. items that have come into contact with small amounts of short-lived radioactive material. In particular, a large percentage of the waste generated during decommissioning and dismantling of a nuclear power plant is LLW. With respect to VLLW, low-level wastes may have either a higher activity due to short-lived radionuclides or a higher content of long-lived radionuclides, therefore they need to be contained in an appropriate infrastructure.

Due to its low activity and correspondingly low radiation doses, LLW can normally be handled using rubber gloves and without particular shielding. In terms of disposal, since the content of long-lived radionuclides is low and there is no significant heat generated by the radioactivity, no long term and secluded repositories are necessary. Therefore, surface or near-surface repositories can guarantee the necessary safety in the disposal of LLW. Near-surface repositories are disposal sites located at ground level, with protecting covering of the order of a few metres thick (see Sect. 6.4), or in caverns below ground level at depths of tens of metres. If a dedicated surface repository is not available, temporary storage is allowed, which can also take place at the same site where the waste was produced, in which case the site operator will be responsible for proper storage and surveillance.

Low-level waste comprises 90 % of the volume but only 1 % of the radioactivity of all radioactive wastes generated in the world each year [3, 4].

6.1.3 *Intermediate-Level Waste (ILW)*

Intermediate-level wastes typically arise from industrial processes, for example they include residues from reactor water treatment and filters used for purifying a reactor's cooling water. ILW typically generates negligible heat and its radioactivity ranges from just above natural background levels to more elevated radioactivity in certain cases, such as for parts coming from inside the reactor vessel in a nuclear power plant or for the non-dissolved metal fuel casings coming from reprocessing of spent nuclear fuel (see below HLW treatment). Depending on the activity level, handling ILW may require to use appropriate shielding.

Specific chemical treatments are typically applied to the ILW to separate as much as possible the long-lived radionuclides from the rest of the material. Once such separation is performed, the part that does not contain significant quantities of long-lived radionuclides can be reclassified into LLW and discharged, provided that its radioactivity is below the limits established by the safety and regulatory authority. If such LLW is in liquid or gaseous form, it can be discharged into river and sea waters or through specific exhaust stacks, respectively.

Instead, the ILW part where the long-lived radioactive constituents have been concentrated can be packaged in a suitable matrix, like concrete, bitumen, or resin. Once properly packaged and depending on the specific long-lived radionuclides contained in it, the treated ILW can be disposed of either in surface, near-surface, or geological repositories.

Worldwide, ILW makes up 7 % of the volume and has 4 % of the radioactivity of all radioactive wastes generated each year [3, 4]. By definition, its radioactive decay generates less than about 2 kW/m^3 of thermal power [2]; therefore for ILW it is not necessary to take into account thermal effects in the design of its storage or disposal facilities.

6.1.4 High-Level Waste (HLW)

High-level waste (HLW) is characterized by a relatively high content of long-lived radionuclides, so that not only it remains radioactive for long time (from hundreds to hundred-thousands years), but also its activity remains high for long time and heat production is substantial. To be more quantitative, in HLW thermal power generation is greater than 2 kW/m^3 [2], and the typical activity levels are in the range of 5×10^{16} to 5×10^{17} Bq/m^3 [2]. Consequently, HLW requires both shielding and cooling.

HLW accounts for over 95 % of the total radioactivity produced in the process of electricity generation [3, 4]. A typical example of HLW is the spent fuel discharged from a reactor core at discharge.

Some countries also reprocess spent fuel to separate and recycle uranium and plutonium, in order to increase the energy produced per unit mass of fuel. However, if recycling is performed in a thermal reactor this process gives rise to additional types of HLW, rich in minor actinides (neptunium, americium, curium, etc.) and fission products. The development and deployment of fast reactors would reduce the production of the latter type of HLW (see Chap. 4). All of this HLW, be it spent fuel or other radioactive material coming from reprocessing, poses a high enough radiological risk to impose a high degree of isolation from the biosphere for a long period of time. Therefore, the disposal of HLW requires deep underground, geological repositories. A typical treatment which is adopted for HLW generated by the reprocessing of spent fuel is vitrification. In this process, HLW is vitrified into borosilicate glass (Pirex), then encapsulated into heavy stainless steel cylinders about 1.3 m high and stored for ultimate deep underground disposal [5].

6.2 Composition of Spent Fuel

As seen in previous Chapters, the spent fuel is the material resulting from a certain period of irradiation of fresh fuel in a nuclear reactor, during which fission and neutron capture processes consume some isotopes and create new ones.

At the end of its life in the reactor, the fuel needs to be characterised accurately for safety reasons. Spent fuel isotopic composition varies as a function of initial composition (kind and amount of fissile and fertile atoms), neutron spectrum, flux, burnup (see Sect. 3.12), design of the fuel elements, position occupied in the reactor during operation, and finally the cooling time after removal from the reactor. Because of this, spent fuel elements from various reactors differ in composition from each other and even between fuel batches from the same reactor.

The management of the spent fuel from nuclear power plants is different according to the different policies, perspectives and fuel cycle scenarios in different countries. However there are important common goals—the need of a deep geological repository, the requirement to reduce the burden of the radiotoxic elements to be disposed of, non-proliferation issues, etc.—which motivated the development of international joint research programmes.

As shown in Fig. 4.10, the spent nuclear fuel contains per kilogram some 35 grams of fission products (^{90}Sr, ^{137}Cs, ^{93}Zr, ^{99}Tc, ^{85}Kr, etc.), 9 grams of various Pu isotopes (about 65 % of which is still fissile) and less than 1 g of minor actinides (Am, Np, Cm, etc.), the balance being uranium. Therefore the actual nuclear waste which needs to be treated and disposed of represents only a tiny fraction of the spent fuel discharged from a LWR (some 36 g per kilogram spent fuel, that is fission products and minor actinides), and the material which might be reused through recycling is more than 95 %.

Nearly all of the fission products are radioactive, some decaying very fast and thus radiating very strongly, others decaying more slowly. No fission products have a half-life in the range 100–210,000 years, nor beyond 15.7×10^6 years.

Fission products with a significant yield and a short half-life (until several days) are responsible for the large amount of decay heat that is produced within the core immediately after reactor shutdown, i.e. after stopping the chain reaction. However, their radioactivity dies away in a few days or weeks, which means that, in case of accidental release of radioactive products, they pose a significant hazard only for such a limited amount of time and that, when the spent fuel is stored in the cooling pools, they do not produce any more heat after a short amount of time.

Other short-lived fission products (usually abbreviated as SLFP) with half-lives until a few years can instead pose a safety concern for a correspondingly long period of time. Similarly for medium-lived fission products (abbreviated as MLFP), with half-lives until about 100 years (e.g. ^{137}Cs and ^{90}Sr, with half-life of about 30 years and ^{151}Sm with half-life of about 90 years). Such isotopes contribute to heat production in the spent fuel and can be a safety concern for decades, if accidentally released into the environment. Finally, seven long-lived fission products (abbreviated as LLFP) have half-lives of 211,100 years (^{99}Tc) and more (^{126}Sn, ^{79}Se, ^{93}Zr, ^{135}Cs, ^{107}Pd and ^{129}I).

After ^{137}Cs and ^{90}Sr have decayed to low levels, the bulk of radioactivity from spent fuel comes not from fission products but from actinides, notably ^{239}Pu (half-life 24,110 a), ^{240}Pu (6561 a) and ^{241}Am (432.2 a).

Figure 6.1 shows the time dependence of specific radioactivity of the major radioisotopes in high-level radioactive waste, which together are responsible for the long-term waste activity. Note that both scales are logarithmic and that the specific activity is measured in Bq per kg of waste from a uranium fuelled reactor.

The isotopes of Fig. 6.1 are important in determining the conditions for disposal of HLW. Their half-lives are given in Table 6.1. As said above, those with short

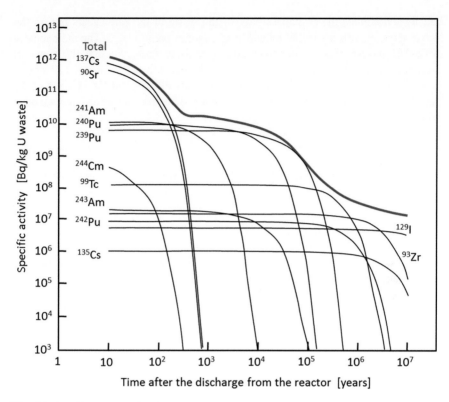

Fig. 6.1 Specific activity of major isotopes in high-level radioactive waste and its decay over time. The top red line shows the total radioactivity. This way one can easily see which isotopes are contributing the most to the total radioactivity of the waste over long time scales

Table 6.1 Radioisotopes with very different half-lives that are significant at different stages of high level waste management and disposal [6]

Isotope	Half-life (a)
Curium-244	18.1
Strontium-90	28.79
Caesium-137	30.0
Americium-241	432.2
Plutonium-240	6561
Americium-243	7370
Plutonium-239	24,110
Technetium-99	213,000
Plutonium-242	373,500
Zirconium-93	1.53×10^6
Caesium-135	2.3×10^6
Iodine-129	1.57×10^7

half-lives dominate the overall activity of the waste in the early years but, in the longer term, the less active but longer-lived isotopes predominate. The radioactivity of Cs-137 and Sr-90 dominates the total radioactivity during the first 100 years. In the longer term, the long-lived isotopes Tc-99, Cs-135, Zr-93 and I-129 are of major concern, as well as the transuranic elements Pu-239 and Pu-242, which by that time are responsible for most of the heat production due to their energetic α decay.

6.3 Amounts of Radioactive Waste Generated by Nuclear Power Plants

In countries where the spent fuel is not reprocessed to separate the plutonium (i.e. countries adopting the so-called once-through fuel cycle), the whole spent fuel is considered nuclear waste. An average nuclear power plant (1 GW(e)) unloads close to 30 tonnes of spent fuel a year, containing about one tonne of true high-level waste (fission products plus minor actinides). During a year the reactor, operating at 80 % capacity, would generate some 7 billion kWh of electricity. Then, considering the whole spent fuel as waste, the waste production of this reactor would be only about 4 mg/kWh. This quantity is much smaller than the waste produced by other conventional sources of electricity production because of the high energy density of nuclear energy (see Sect. 3.3 and Problem 3.4). This tiny amount of waste can be further reduced if the spent fuel is recycled. In the closed fuel cycle, uranium and plutonium are separated from the spent fuel and recycled to fabricate mixed-oxide fuel (commonly referred to as MOX), which is further burnt to produce electricity. Therefore, in this type of fuel cycle, plutonium is not disposed of as waste as it is the case in the once-through fuel cycle (this aspect will be discussed in more detail in Sect. 6.6). In this case, the waste comprises only minor actinides and fission products. Therefore, considering again a typical nuclear power plant of 1 GW(e), after fuel reprocessing the nuclear waste production would be 0.14 mg/kWh (1 tonne of true high-level waste out of 30).

Both the non-reprocessed spent fuel and the separated waste containing minor actinides and fission products belong to the category of high-level waste, due to their high activity and high content of long-lived radionuclides.

Over the years, there has been a general trend towards a reduction in the volume of waste generated per unit of electricity produced through improved practices and technologies [7, 8]. Table 6.2 gives indicative volumes of radioactive waste produced annually by a typical 1000 MW(e) nuclear plant, for once-through cycle and with reprocessing of the spent nuclear fuel (SNF). The closed fuel cycle gives a HLW volume reduction by a factor 1.5–3 (remember that SNF is in fact HLW).

The production rate of waste of all levels from electricity generation from nuclear power, including the portion coming from fuel fabrication facilities and from plant decommissioning and dismantling, is about 4×10^5 t/a, corresponding

Table 6.2 Indicative volumes (in m^3) of radioactive waste produced annually by a typical 1000 MW(e) nuclear plant, for once-through cycle and with reprocessing of spent nuclear fuel (SNF) [9]

Waste type	Once-through fuel cycle	Closed fuel cycle
LLW/ILW	50–100 m^3	70–190 m^3
HLW	0 m^3	15–35 m^3
SNF	45–55 m^3	0 m^3

Table 6.3 Solid waste production rate for electricity generation from coal fired plants and nuclear plants

	Waste production rate	
	[t/a]	[kt/TWh]
Coal-fired plant	6×10^8 ash	90 ash
Coal mining and related activities	2×10^{10}	3000
Nuclear power plant	4×10^{5a}	0.2^a
Mining and uranium milling	4.5×10^7	8

[a]Including waste originating from eventual plant decommissioning and dismantling

to a volume of roughly about 200,000 m^3. Within this waste, only about 2.5 % (i.e. only about 10,000 t/a), is high-level waste (HLW) or spent nuclear fuel (SNF), which, on the other hand, account for most of the associated radioactivity. Waste production rate from supplying raw materials (mining and uranium milling) is less than 4.5×10^7 t/a [2].

It is interesting to compare the waste production in electricity generation from coal-fired power plants with that from nuclear power plants. According to Ref. [2], the waste production rate of coal-fired plants is about 6×10^8 t/a of ash and 10.5×10^9 t/a of CO_2, and that from coal mining and related activities is 20×10^9 t/a. These values are given in Table 6.3, together with the corresponding values from nuclear power plants. For a more significant comparison, in the table the amounts of waste normalized to the produced energy are also given.

The Table shows that the absolute amount of solid waste from nuclear electricity production is about 0.07 % of the ashes from the coal sector, and total wastes from mining and related activities from nuclear electricity production are about 0.2 % of those from the coal sector. Larger but still small percentages are obtained from the amounts of waste per unit of energy produced. In addition, as seen in Sect. 4.3, the CO_2 emission per kWh from the nuclear electricity sector is estimated to be less than 2 % of that from coal-fired plants [10].

Yet another useful comparison is with the worldwide production of any kind of waste, including non-radioactive waste, which is the vast majority. According to Ref. [2], the global waste production rate is in the order of $8–10 \times 10^9$ t/a (without counting wastes from mining and related activities), of which about 4×10^8 t/a is hazardous waste,[1] and about 4×10^5 t/a is radioactive waste from nuclear power plants and their fuel cycle support facilities (excluding mining and extraction wastes). In comparison with industrial toxic and hazardous waste, the volume of radioactive waste from nuclear power generation is therefore relatively small. If we considered all nuclear waste as hazardous (which is a very conservative assumption as a large percentage of it is VLLW and LLW), it would be about 0.1 % of the worldwide production of hazardous waste.

As a final remark, it should be noted that limited quantities of radioactive waste are also generated by industries and hospitals, and need to be properly stored or disposed of as well.

Problem 6.1: Mass of charged fuel. Consider a reactor with 1000 MW(e) power, fuelled by uranium dioxide (UO_2) fuel at 4 % enrichment. Assume that, after producing $E = 2.1 \times 10^{10}$ kWh, the enrichment has decreased to 1 %. Calculate the total amount M_O of uranium oxide present in the reactor ($^{235}UO_2 + {}^{238}UO_2$) at the beginning of the cycle, assuming that each fission releases 200 MeV of energy.
Solution: Recalling that 200 MeV = 3.2×10^{-11} J of energy, in order to produce the energy

$$E = 2.1 \times 10^{10} \, \text{kWh} = 2.1 \times 10^{13} \times 3600 \, \text{J} = 7.56 \times 10^{16} \, \text{J}$$

one has to fission

$$N = \frac{E}{3.2 \times 10^{-11}} = \frac{7.56 \times 10^{16}}{3.2 \times 10^{-11}} = 2.36 \times 10^{27} \text{ nuclei U-235.}$$

Calling $N_A = 6.022 \times 10^{23}$ the Avogadro's number and assuming that the atomic weight of ^{235}U is $P_U = 0.235$ kg, the mass of ^{235}U that has undergone fission is

$$m = N\frac{P_U}{N_A} = 2.36 \times 10^{27} \frac{0.235}{6.022 \times 10^{23}} = 921 \, \text{kg.}$$

Calling M_1 and $M_2 = (M_1 - m)$ the masses of ^{235}U in the fresh and spent fuel, respectively, the total mass M of uranium ($^{235}U + {}^{238}U$) in the fuel can be easily obtained observing that

---

[1]Hazardous waste is a waste with properties that make it potentially dangerous or harmful to human health or the environment. The universe of hazardous wastes is large and diverse: they can be liquids, solids, or contained gases, and can be the by-products of manufacturing processes, discarded used materials, or discarded unused commercial products, such as cleaning fluids (solvents) or pesticides. In regulatory terms, a hazardous waste is a waste that exhibits one of the following four characteristics—ignitability, corrosivity, reactivity, or toxicity.

$$M_1 = 0.04M, \quad \text{and} \quad M_2 = M_1 - m = 0.01(M - m),$$

which gives:

$$0.04M - m = 0.01(M - m)$$

and

$$M = \frac{0.99}{0.03} m = 33\, m = 30{,}393\, \text{kg}.$$

The molecule of UO_2 contains the fraction f of uranium (see Problem 4.1)

$$f = \frac{0.04 \times 235 + 0.96 \times 238}{0.04 \times 235 + 0.96 \times 23 + 2 \times 16} = 0.88\,.$$

Then the fuel charged in the reactor has mass

$$M_O = \frac{M}{f} = \frac{30{,}393}{0.88} = 34.5\, \text{tonnes}.$$

Problem 6.2: Volume of spent fuel. If all fuel in the reactor of the Problem 6.1 were discharged after producing the amount E of electricity, this would be the total amount of spent fuel produced in this reactor cycle. Calculate the volume V occupied by such spent fuel (density of the uranium dioxide is $d = 10.97$ g/cm^3) and the amount in kg of fission products contained in it (neglect the mass converted into energy).
Solution: The volume of the spent fuel is

$$V = \frac{M_O}{d} = \frac{3.45 \times 10^4\, \text{kg}}{10.97 \times 10^3\, \text{kg/m}^3} = 3.14\, \text{m}^3.$$

Neglecting the mass converted into energy, all the mass $m = 920$ kg of the ^{235}U that undergoes fission is converted into fission products.

Problem 6.3: Decay heat from α-emitter. Assume that the spent fuel of Problem 6.2 contains 30 kg of ^{241}Am, which undergoes α decay with a half-life of 432.2 years and an energy released per decay $E = 5.64$ MeV. Calculate: (a) the rate of total thermal energy production by such radioactive decay and the rate of thermal energy production per unit volume (assume the americium to be uniformly spread within the spent fuel); (b) the total heat produced by such a radioactive decay in watt and the heat per unit volume, at discharge from the reactor, in W/cm^3.

Solution:

(a) From Eqs. (2.9) and (2.12) the activity R of the given amount of ^{241}Am is $R = \lambda N_0 e^{-\lambda t}$, where N_0 is the number of ^{241}Am nuclei contained in the mass $M = 30$ kg and λ is the decay constant of ^{241}Am. Calling $N_A = 6.022 \times 10^{23}$ the Avogadro's number and assuming that the atomic weight of ^{241}Am is $P_A = 0.241$ kg, one has

$$N_0 = \frac{M}{P_A} N_A = \frac{30}{0.241} \times 6.022 \times 10^{23} \approx 7.5 \times 10^{25} \text{ nuclei.}$$

Recalling that 1 year $\approx 3.15 \times 10^7$ s, one has

$$\lambda = \frac{0.693}{\tau_{1/2}} = \frac{0.693}{432.2 \times 3.15 \times 10^7} = 5.1 \times 10^{-11} \text{ s}^{-1}.$$

Then the rate of heat production is

$$P = ER = E\lambda N_0 e^{-\lambda t} = 5.64 \times 5.1 \times 10^{-11} \times 7.5 \times 10^{25} e^{5.1 \times 10^{-11}t}$$
$$= 2.16 \times 10^{16} e^{5.1 \times 10^{-11}t} \text{ MeV/s.}$$

(b) At $t = 0$, recalling that 1 MeV $= 1.6 \times 10^{-13}$ J, we have that

$$P = 2.16 \times 10^{16} \times 1.6 \times 10^{-13} = 3.46 \times 10^3 \text{ W} = 3.46 \text{ kW.}$$

With the volume of 3.14 m^3 = 3.14 $\times 10^6$ cm^3 obtained in Problem 6.2, this corresponds to 1.1×10^{-3} W/cm^3.

Problem 6.4: Transport of fuel. Compare two 500 MW electric power stations, one burning oil and the other using 3.0 % enriched uranium in the ^{235}U isotope. Both stations operate 6000 h/a at 35 % thermal-to-electricity efficiency. The oil (43.5 MJ/kg combustion energy) is carried by 100,000 tonne carrying capacity oil tankers, and the uranium fuel by train cars of 20 t capacity each. Calculate (a) how many oil tankers will be needed every year for the oil-fired station, (b) How many train cars will be needed every year for the nuclear power station for transporting the enriched UO$_2$ reactor fuel? Assume a reactor fuel rating (i.e. energy per mass unit) of 40 MW(th)d/kgU (1 MWd = 1 MW multiplied by one day = 24 MWh = 24,000 kWh, here U refers to the uranium fuel, typically containing 3 % of ^{235}U).

Solution:

(a) The total energy produced by the plants in 6000 h of operation is

$$E = \frac{500 \times 10^6 \times 6000 \times 3600}{0.35} = 3.09 \times 10^{16}\,\text{J}.$$

To produce this energy the oil plant burns a mass

$$M_\text{o} = \frac{3.09 \times 10^{16}}{43.5 \times 10^6} = 7.1 \times 10^8\,\text{kg}.$$

Therefore seven oil tankers (carrying capacity 10^8 kg) will be necessary $(7.1 \times 10^8 / 1 \times 10^8 = 7.1)$ each year.

(b) To produce the energy E, the nuclear plant burns a mass of ^{235}U given by

$$M_{\text{U}-235} = \frac{3.09 \times 10^{16}}{40 \times 10^6 \times 3600 \times 24} 0.03 = 268\,\text{kg}.$$

The total uranium mass in the fuel is

$$M_\text{U} = \frac{M_{\text{U}-235}}{0.03} = 8941\,\text{kg}.$$

The molecule of UO_2 contains the fraction f of uranium (see Problem 4.1)

$$f = \frac{0.03 \times 235 + 0.97 \times 238}{0.03 \times 235 + 0.97 \times 23 + 2 \times 16} = 0.87.$$

Then, the mass of uranium dioxide charged in the reactor is

$$M = \frac{M_U}{0.87} = 10{,}277\,\text{kg}.$$

To transport this mass, half train car of 20 t capacity ($10.3/20 \approx 0.5$) is needed each year.

6.4 Radioactive Waste Disposal

The world has over half a century of knowledge and experience on how to deal with nuclear waste. Good practices developed over the years are being used throughout the whole cycle of electricity production from nuclear power to help ensure the safety of people and the environment from possible hazardous effects of radiation.

An international framework was set in 1995 by the IAEA, by defining a set of internationally agreed fundamental principles for the safe management of radioactive waste [11]: protection of human health, protection of the environment, protection beyond national borders, protection of future generations, burdens on future generations, national legal framework, control of radioactive waste generation, radioactive waste generation and management interdependencies, and finally, safety of facilities.

The characteristics of nuclear waste are well known. This is a prerequisite for safe and secure disposal, which is the final step in radioactive waste management. The appropriate disposal options and the extent of safety measures depend on the length of time the waste remains hazardous—some waste remains radioactive for hundreds of thousands of years and other waste for tens of years or less.

Disposal of low-level waste and intermediate-level waste is already implemented in several countries. Usually the disposal facilities are at, or near, the surface, but some ILW that contains long-lived radioactivity requires disposal at greater depths, of the order of tens of metres to a few hundred metres, which is currently being studied [2]. Before storage or disposal, LLW and ILW often undergo physical and chemical treatments called *conditioning*. The purpose of such treatments is to transform the waste into a solid form offering robustness and stability, so that it can be handled, stored, transported, and disposed of (in the case of long-lived waste). The final product of the conditioning process is a solidified matrix (e.g. concrete or glass), that encapsulates the radioactive material, in turn placed inside a suitable container (e.g. a steel barrel). Such conditioned products, often called *waste packages*, can be further placed in additional containment structures, then stored in a surface facility or disposed in a near-surface facility (see Fig. 6.2). In this form of disposal, the combination of engineered barriers provided by conditioning form, placement within disposal units and construction of a cover, allow to keep the waste confined for periods of time in the order of 300–500 years [2].

Spent nuclear fuel and other HLW are presently temporarily stored in cooling ponds in storage facilities, often hosted by the same sites where such waste was produced. This temporary storage phase is an important step in the safe management of radioactive waste, as it helps to reduce both radiation and heat generation prior to waste handling and transfer to the final disposal site. Experience matured over past decades has demonstrated that, as long as active surveillance and maintenance are ensured, the interim storage of radioactive waste can be safe. Moreover, the storage is technically feasible and is a safe solution for several decades if monitoring, control and care are properly implemented.

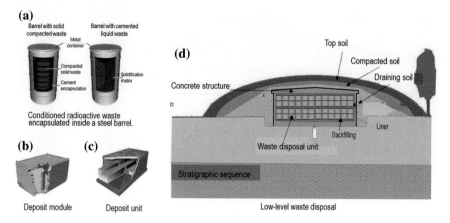

Fig. 6.2 Conditioning and storage of LLW and ILW; **a** conditioning methods for solid (*left*) and liquid (*right*) waste; **b** further accommodation of barrels within modules; **c** further accommodation of modules within units; **d** final near-surface infrastructure hosting all units. Adapted from [12]

As far as final disposal is concerned, several options are being examined and research on how to implement disposal is being conducted in many countries with operating nuclear power plants. In every option, deep geological disposal is the preferred final end point (see Fig. 6.3). This is a technically viable and safe method for isolating the spent nuclear fuel, as well as the high-level and long-lived radioactive waste.

The principle of geological disposal is to isolate the waste deep inside a suitable host formation, e.g. granite, salt or clay. The waste is placed in an underground disposal facility, designed to ensure that a system of natural and multiple artificial barriers work together to prevent radioactivity from escaping (a principle resembling the defence-in-depth concept applied to nuclear power plants, see Sect. 5.3). The multiple artificial barrier is an engineered barrier system which comprises the solid waste matrix (i.e. the material in which the waste is immersed, such as concrete or resin) and the various containers and backfills used to immobilize the waste inside the repository. The natural barrier (i.e. the geosphere) is principally the rock and groundwater system that isolates the repository and the engineered barrier system from the biosphere. The host rock is the part of the natural barrier in which the repository is located. Emplacement of the waste in carefully engineered structures placed at depth in suitable rock is chosen principally for the long-term reliability of containment that the geological environment provides. At depths of several hundred metres in a tectonically stable environment, processes that could disrupt the repository are so slow that it is expected that the rock and groundwater systems will remain almost unchanged for hundreds of thousands of years, and possibly longer.

The ultimate goal in this approach to the disposal of the most hazardous and long-lived waste is to confine it for sufficiently long time (up to hundreds of thousands of years) in a system naturally offering safety features so that there would

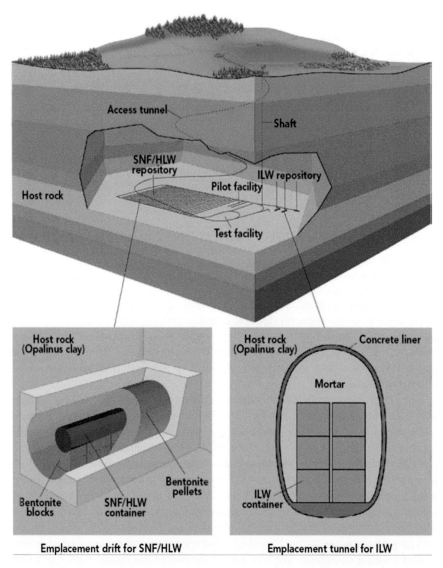

Fig. 6.3 Schematic concept of a multiple-barrier system for geological disposal of spent fuel and high-level waste (SNF/HLW). The waste is loaded into specially designed corrosion-resistant containers that are placed in horizontal, or vertical boreholes drilled several hundred metres down into the rock and sealed with bentonite clay. The SNF/HLW is monitored and retrievable until the site is closed, at which point the access tunnel is backfilled and sealed. In the figure, a space for intermediate-level waste (ILW) is also indicated [13]

not be any significant spread of radioactivity into the biosphere at the Earth's surface. Since safety would be guaranteed in a passive way, i.e. without intervention of engineered systems operated by humans, there would not be any burden on future generations in terms of continuing surveillance of the site. However, a challenging aspect in implementing geological disposal is to convince the public that the accumulated knowledge on geology and involved materials, as well as extended tests in pilot facilities, are sufficient to ensure the desired long-term confinement.

Even though natural barriers are thought to provide an essential part of the waste confinement, also engineered barriers are needed to provide further containment in addition to natural ones. Therefore, HLW has to undergo conditioning before any transportation, storage or disposal. In this case, conditioning typically comprises incorporation at high temperature inside a steel container (*flask*), sealed by welding. The flask is then placed inside cylindrical tanks (*casks*), typically with 2.5 m diameter and 4.5 m height, suitable for transportation and storage within long-term facilities, while final disposal is prepared.

No deep underground repository for HLW (including spent fuel) has yet started operation, with the exception of a facility for disposal of non-heat generating HLW, which is in operation in the USA since 1999. It is the Waste Isolation Pilot Plant (WIPP), near Carlsbad in the New Mexico [14], which is used to store transuranic waste left over from nuclear weapons research and testing operations from past defence activities in the USA. The facility has disposal rooms mined about 700 m down a saline formation that has been stable for more than 200 million years, and has required 25 years for its construction.

Programmes to dispose of spent fuel have been studied extensively in several countries. Site characterization and selection for deep geological repositories have been under way since the 1970s. Identification of a suitable geological site requires a long-term project that is not within the reach of all countries. This is why the European Union discussed for some time the possibility of identifying a common geological site, with the purpose to serve several countries. In fact, the limited volume of HLW and the uncertainty about the aspect of the possible retrievability or not of the waste forms (which is an especially sensitive issue when the material to be disposed of is spent fuel, which in principle may be reprocessed in the future for recycling) appear to be the aspects that have delayed the decision process in some countries. However, it must also be considered that decisions in this delicate matter have to wait for the results of research and public discussions on the subject.

The two countries closest to licensing and start-up of operation of a deep geo-logical disposal facility for civil waste are Finland and Sweden. France is currently preparing a license application for geological disposal in a clay rock host formation. These projects are currently the most advanced disposal programmes in the world. They are scheduled to start operating between 2020 and 2030.

Storage and disposal are complementary and not competing activities, and both are needed to ensure safe and reliable radioactive waste management. Waste storage is a step in the management strategy leading to final disposal. When a disposal site is available, ILW and HLW can be sent there directly at regular intervals. If not, interim storage in an appropriate containment structure above ground is necessary. For HLW and spent fuel, it has always been recognised that interim storage to permit decay of radiation and of heat generation is necessary. The timing and duration of these options depend on many factors. Perpetual storage in these engineered repositories is not feasible because active controls cannot be guaranteed forever, but there is no urgency for abandoning interim storage on technological or economic grounds. However, ethical reasons and safety considerations require the establishment of final disposal facilities in due course.

In the nuclear energy sector, good waste management resulting in safe disposal also considers financial implications. The objective is to have enough funds to cover all aspects of activities required once electricity production of a facility has ceased (i.e. waste disposal, decommissioning, human resources, etc.). There are mechanisms for collecting money to cover all nuclear power production expenses: for instance in some countries the companies operating the plant have to secure a certain amount of money out of their revenue, to guarantee the necessary funds for decommissioning, storage and disposal. In terms of good practices of radioactive waste management, responsibility covers all the steps from *cradle to grave*, i.e. from uranium mining up to the disposal of the waste.

6.5 The Oklo Natural Fission Reactors

There is an interesting example of an underground long-term storage of radioactive waste that is found in nature, which gives some indication that safe long-term geological storage may be feasible. Two billion years ago, in what is now Oklo in Gabon, West Africa, at least 17 natural nuclear reactors spontaneously achieved criticality in concentrated deposits of uranium ore [15]. At that time, the concentration of ^{235}U in all natural uranium was about 3 % instead of the present 0.720 %. As seen in Sect. 3.8.1, 3 % is roughly the level provided artificially in enriched uranium used to fuel most nuclear power plants.

The reactors were discovered in 1972, when assays of mined uranium showed only 0.717 % U-235 instead of 0.720 % as everywhere else on the Earth, on the moon and even on meteorites. Further investigation identified significant concentrations of fission products from both uranium and plutonium.

The natural chain reactions at Oklo started spontaneously because of the presence of groundwater acting as a moderator, continued overall for about 2 million years before finally dying away. Each reactor operated in pulses of about 30 min— interrupted when the water turned to steam so ceasing to function as moderator and, thereby, switching the reactor off for a few hours until it cooled. It is estimated that about 130 TWh of thermal energy were produced.

During the long reaction period about 5.4 tonnes of fission products as well as up to 2 tonnes of plutonium together with other transuranic elements were generated in the ore body. The initial radioactive products have since long decayed into stable elements but close study of the amount and location of these has shown that there was little movement (a few centimetres) of radioactive wastes during and after the nuclear reactions. Plutonium and the other transuranics remained immobile.

The natural fission reactors at Oklo can be considered as analogues for very old radioactive waste repositories and can be used to study the transport behaviour of transuranic nuclides and the stability of uranium minerals that have undergone criticality.

Problem 6.5: U-235 percentage in a uranium ore. The natural nuclear fission reactors at Oklo, in the African nation of Gabon were discovered when assays of mined uranium showed only 0.717 % U-235 instead of 0.720 % as everywhere else on the Earth's crust. Calculate in how much time T from now the percentage of U-235 in uranium ore will decrease to 0.717 % (use the values of the half-lives of the uranium isotopes given in Table 2.2).

Solution: Let us call N_0, N_1 and N_2 the number of all initial uranium nuclei in a typical ore and those of nuclei U-235 and U-238 remaining at the time T, respectively.

Using the radioactive decay law of Eq. (2.9) we have

$$N_1 = \alpha N_0 e^{-\lambda_1 T},$$
$$N_2 = (1 - \alpha) N_0 e^{-\lambda_2 T},$$

where $\alpha = 0.720$ % is the percentage of U-235 nuclei in the ore at time $t = 0$. By dividing the first equation by the second, one obtains:

$$\frac{N_1}{N_2} = \frac{\alpha}{(1 - \alpha)} e^{-(\lambda_1 - \lambda_2) T}.$$

At the time T the percentage of U-235 nuclei is $\beta = 0.717$ %, that is $N_1 = \beta(N_1 + N_2)$. Then the previous equation can be written

$$\frac{\beta}{(1 - \beta)} = \frac{\alpha}{(1 - \alpha)} e^{-(\lambda_1 - \lambda_2) T},$$

Taking the log of both sides and recalling that decay constant and half-life are connected by Eq. (2.10) ($\lambda = 0.693/\tau_{1/2}$) gives:

$$
T = \frac{1}{\lambda_2 - \lambda_1} \ln \frac{(1 - \alpha)\beta}{\alpha(1 - \beta)}
$$
$$
= \frac{4.47 \times 10^9 \times 7.04 \times 10^8}{0.693 \times (7.04 \times 10^8 - 4.47 \times 10^9)} \ln \frac{(1 - 0.720 \times 10^{-2}) \times 0.717 \times 10^{-2}}{0.720 \times 10^{-2}(1 - 0.717 \times 10^{-2})} = 5.07 \times 10^6 \, \text{a}.
$$

It is a huge lapse of time, which justifies the alarm raised by the tiny discrepancy observed in the Oklo ore.

6.6   Research on Partitioning and Transmutation

As discussed in the previous sections, the most relevant hazard from the spent fuel
of nuclear power plants is given by long-lived heavy nuclides (plutonium and minor
actinides) and long-lived fission products, which require robust isolation from the
environment for extremely long time.

A measure of the potential biological hazard of a radionuclide is provided by its
radiotoxicity. It is the product of the activity of that nuclide and the effective dose
coefficient $e(T)$, which accounts for radiation and tissue weighing factors, as well as
metabolic and biokinetic information (absorption in the organism, residence time in
the body, etc.). Radiotoxicity is commonly expressed in sieverts. The quantity T is
the integration time in years following intake. For adults $T = 50$ years and it is
assumed that the radioisotopes are ingested or inhaled. The ingested radiotoxicity
for a given isotope is determined by the activity (Bq) multiplied by the isotope
effective dose coefficient (measured in Sv/Bq), for ingestion. The inhaled
radioactivity is defined similarly, but with a different coefficient, for inhalation.

Studies are currently underway in many countries, with the aim of further
reducing the radiotoxicity of the nuclear waste. In principle, this goal may be
achieved by separating (*partitioning*) the long-lived actinides (neptunium, plutonium, americium, curium, etc.) and transforming (*transmuting*) them into elements
with average shorter lifetime or non-radioactive elements, thereby eliminating or
reducing the radiological hazard and waste disposal problem. Such transmutation
can be performed by means of neutron irradiation in critical or subcritical reactors
driven by particle accelerators (see below). In the case of subcritical reactors, for
instance, a proton accelerator would serve the purpose of creating an intense
neutron source, so that the neutron chain reaction in the subcritical reactor could go
on thanks to such an external neutron source.

Among the fission products, for example technetium-99 and iodine-129 are very
long-lived (half-lives of 2.1×10^5 a for decay to stable ruthenium-99 and
15.7×10^6 a for decay to stable xenon-129, respectively) and therefore should
undergo geological disposal given the time needed for their radioactivity to decay
significantly. At the same time, their isolation from the environment is difficult
because both of them are highly soluble in groundwater, in which case they can
present a high mobility. By exposing these isotopes to a neutron flux, both of them
can be transmuted via radiative neutron capture and subsequent decay through the
reactions

$$n + {}^{99}\text{Tc}(2.1 \times 10^5 \text{ a}) \rightarrow {}^{100}\text{Tc}(\beta^--\text{decay}, \tau_{1/2} = 16 \text{ s}) \rightarrow {}^{100}\text{Ru}.$$

$$n + {}^{129}\text{I} \rightarrow {}^{130}\text{I} \rightarrow (\beta^--\text{decay}, \tau_{1/2} = 12.36 \text{ h}) \rightarrow {}^{130}\text{Xe}.$$

Among the radioactive wastes produced in nuclear reactors, transuranic actinides
also represent an issue. The latter comprise plutonium (which however can be
recycled as fuel) and minor actinides (mainly neptunium, americium, and curium),
some isotopes of which have half-lives of hundreds of years to millions of years.

Some of these transuranics are fissile and many of them are fissionable, i.e. can be fissioned by fast neutrons. Since many fission products are relatively short-lived (many have half-lives of less than 100 years), by fissioning actinides the long-term hazard is reduced to a shorter-term one.

By applying specific radiochemical processes, it is possible to separate the plutonium from the minor actinides and from the fission products. Then the fission products should be exposed to a soft (thermal) neutron spectrum, for which typically the capture cross section is highest, while the minor actinides should be exposed to a fast neutron spectrum, which will burn of all them by fission. However, large quantities of minor actinides cannot be loaded into a critical reactor, as the percentage of their delayed neutrons, fundamental for reactor control (see Sect. 3.10), is smaller than for uranium and plutonium. For dealing with large quantities of minor actinides one should instead consider a subcritical core ($k_{eff} < 1$), coupled to an external neutron source that supplies the additional neutrons needed to keep the chain reaction going. An external neutron source can be obtained for instance by absorbing a particle beam from an accelerator into a thick target. In this case, the time behaviour of the system is essentially determined by the neutron source and in case of unwanted reactivity increases, the chain reaction can be quickly shut down by turning off the accelerator. Such a system is called an *Accelerator-Driven System* (ADS) [16].

Transmutation facilities and ADS have been or are being studied in several countries, prototypes built and large-scale systems designed [17, 18], but no actual waste incinerator has been built yet.

The modification of the radiotoxicity due to the main fuel cycle strategies is shown in Fig. 6.4. The red curve shows, in relative units, the time dependence of the radiotoxicity of spent fuel, as discharged from the reactor (uranium + plutonium + minor actinides + fission products). The time requested for the radiotoxicity of this fuel to decrease to the level of the original natural uranium (horizontal black line is the radiotoxicity of the amount of natural uranium that was used to produce the initial amount of nuclear fuel) is about 250,000 years. This is the standard way to evaluate when nuclear waste can be considered no longer dangerous for the environment. This is the situation for most of the existing reactors of Generation II, which adopt the nuclear open fuel cycle (or once-through cycle) without reprocessing.

In the partially closed fuel cycle, where reprocessing is used to extract the plutonium and form new fuel in the form of MOX to be recycled in the core, the decay time of the residual waste (actinides and fission products) is decreased to about 10,000 years (blue curve in the figure). This process is already deployed in some Generation II reactors under operation (e.g. in France), and foreseen in a number of Generation III and Generation III+ reactors, currently under construction.

The green curve shows how the radiotoxicity would be reduced by the recycling and transmutation not only of uranium and plutonium but also of minor actinides into shorter-lived elements. This way the waste, basically composed of fission products only, would essentially become intermediate-level waste (that reaches the

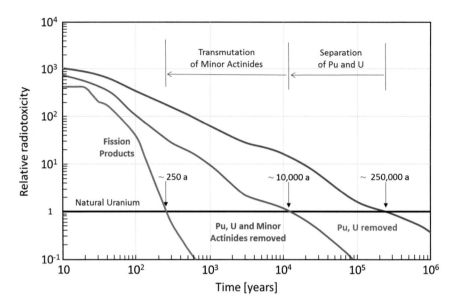

Fig. 6.4 Decay of radioactive materials with high-activity and long half-life contained in spent nuclear fuel. The irradiated fuel discharged from the reactor, which contains uranium, plutonium, minor actinides and fission products, goes back to the original radioactivity of natural uranium from the ore in about 250,000 years (*red curve*). Recycling of uranium and plutonium produces waste containing only minor actinides and fission products, thereby reducing the period of decay by a factor of about 25 (*blue curve*). Fully closed fuel cycles including partitioning and transmutation of the minor actinides would reduce the decay time span by a factor of about 1000 (*green curve*), as the final waste would contain fission products only. Adapted from [19]

natural uranium level in about 250 years), offering an alternative solution for the management of radioactive materials, where geological disposal would only be used for the smaller fraction of long-lived isotopes.

6.6.1   Fast Reactors and Subcritical Reactors Driven by Particle Accelerators

In the effort to improve the safety, security and efficiency of nuclear plants, new concepts of reactors are being developed with ambitious goals including

- the minimization of the production of minor actinides,
- a better and more efficient use of the fuel,
- a better thermodynamic efficiency,
- the possible production of hydrogen at high temperatures,
- improved safety features to minimize the risk of accidents and eliminate the need of an Emergency Planning Zone.

All these various concepts are foreseen in the so-called Generation-IV reactors (see Chap. 4), which were proposed at the beginning of the year 2000 by the Generation IV International Forum (GIF) [20] and are now being developed at national at international level.

Three out of the six reactor concepts endorsed by the GIF are fast reactors. As an example, we have seen in Sect. 3.11 that in a fast reactor it is possible to burn not only fissile elements like ^{235}U and ^{239}Pu, but also fissionable elements like ^{238}U for which fission occurs significantly only above a certain energy threshold around 1.2 MeV (see Fig. 3.4). Although given the much lower fission cross section typically the fuel has to be richer in fissile content, this means that also ^{238}U can be to some extent considered to be a fuel as well. Furthermore in a fast neutron spectrum the thermal fission factor η of Pu is particularly high (see Fig. 3.10), thus allowing breeding of the fuel. These two circumstances have obvious implications in terms of how long uranium resources will last.

Another important aspect in fast neutron reactors is that, for energies, say, above 1.2 MeV, fission becomes dominant over capture, as again can be seen in Fig. 3.4. This means that the production of minor actinides, which occurs namely via neutron capture, is relatively suppressed. Yet another important fact is that among the minor actinides, some are fissionable, i.e. they can undergo fission for neutron energies above about 1.2 MeV. The above considerations imply that in a reactor core with a hard neutron energy spectrum, not only less minor actinides are produced, but those produced can be partly destroyed in the reactor itself. This is why worldwide, currently especially in Europe, Russian Federation and Asia, a number of fast reactor designs are being developed and, in a few cases, also deployed with the final goal to make nuclear power more sustainable in the long term.

6.6.2 Impact of Partitioning and Transmutation on Geological Disposal

Besides the substantial reduction of the potential radiotoxicity source in a repository, the partitioning and transmutation strategy allows in principle a combined reduction of the radionuclide masses to be stored, their associated residual heat, and, as a potential consequence, the volume and the cost of the geological repository.

The impact of partitioning and transmutation strategies on the reduction of radiotoxicity has been the object of numerous well-documented national and international studies.

Generally, the high-level radioactive waste arising from the advanced fuel cycle scenarios, based on fast neutron systems and partitioning and transmutation, generates less heat than the spent fuel. Especially in the case of a fully closed fuel cycle, the considerably smaller thermal output of the high-level waste at the typical considered cooling time in the interim storage (50–100 years) allows for a significant reduction in the total length of disposal galleries required. Separation of caesium and strontium allows for further reduction in the HLW repository size. For example, in the case of disposal in a clay formation the required length of the HLW disposal galleries is reduced by a factor of 3.5 when comparing the waste from the fully closed fuel cycle with the one from a once-through fuel cycle, and by a factor of nine when caesium and strontium are separated.

Extending the cooling time from 50 to 200 years would result in a further reduction of the thermal output of the high-level waste and, consequently, of the required size of the repository. In the case of advanced fuel cycles, this reduction is a factor of about 30.

A large number of different fuel cycles and scenarios have been compared in several national and international studies during the last decade in terms of normalized transuranic element consumption and in terms of transuranic element masses (and associated heat load and radiotoxicity) sent to the waste repository. Such studies considered different fuel cycles adopting fast reactors, light-water reactors and accelerator driven systems in different combinations.

The results of these studies clearly indicate that to reach the optimum performances of partitioning and transmutation, fast spectrum reactors and fully closed fuel cycles are needed, together with chemical processes that allow reaching $\sim 99.9~\%$ recovery of all transuranic elements.

Problem 6.6: Activity of natural uranium. Natural uranium contains 99.275 % of ^{238}U, 0.720 % of ^{235}U, and 0.005 % of ^{234}U. In a uranium ore U_3O_8 (a crystalline blend of UO_2 and UO_3 in 1–2 ratio, known as pitchblende or yellow cake), isotope ^{238}U is in secular equilibrium with 14 other radioactive isotopes (see Fig. 2.2), each having the same activity as the isotope ^{238}U. Similarly, isotope ^{235}U is in secular equilibrium with 11 other radioactive isotopes. Calculate the activity of 1 tonne of natural uranium.

Solution: The activity R of a specific amount of a radionuclide is calculated according to Eq. (2.12) as the decay constant λ multiplied by the number of radionuclides. In the mass $M = 1000$ kg of U_3O_8 the numbers of ^{238}U and ^{235}U nuclei are

$$N_{238} = 3 \times 0.99275 \times N_A \frac{M}{A} = 3 \times 0.99275 \times 6.022 \times 10^{23} \frac{1000}{0.842} = 2.13 \times 10^{27}$$

$$N_{235} = 3 \times 0.00720 \times N_A \frac{M}{A} = 3 \times 0.00720 \times 6.022 \times 10^{23} \frac{1000}{0.842} = 1.54 \times 10^{25}$$

where N_A is the Avogadro number and $A = (3 \times 0.238 + 8 \times 0.016) = 0.842$ kg is the atomic mass of U_3O_8.

Then

$$R_{238} = \lambda_{238} N_{238} = \frac{0.693}{\tau_{238}} N_{238} = \frac{0.693}{4.47 \times 10^9 \times 3.15 \times 10^7} \times 2.13 \times 10^{27}$$
$$= 10.48\,\text{GBq},$$

$$R_{235} = \lambda_{235} N_{235} = \frac{0.693}{\tau_{235}} N_{235} = \frac{0.693}{7.04 \times 10^8 \times 3.15 \times 10^7} \times 1.54 \times 10^{25}$$
$$= 0.48\,\text{GBq}.$$

In the natural pitchblende the two uranium isotopes are in secular equilibrium with all nuclides of their radioactive families, then the activity of ^{238}U has to be multiplied by a factor of 14 (compared to the pure ^{238}U) and activity of ^{235}U by a factor of 11 (compared to pure ^{235}U):

$$R_{\text{U-nat}} = 10.48 \times 10^9 \times 14 + 0.48 \times 10^9 \times 11 \cong 152 \times 10^9\,\text{Bq per tonne}.$$

Problem 6.7: Activity of uranium ore. A low-grade uranium ore contains 0.1 % U_3O_8. Calculate how much ore is needed to obtain 1 kg of uranium oxide UO_2 and how much is needed to obtain 1 kg of uranium oxide UO_2 enriched to 3 % in ^{235}U. Calculate the total activity of the latter quantity of uranium ore and its specific activity (i.e. activity per unit mass). (Assume that all uranium has atomic mass 238).

Solution: In the mass $M = 1$ kg of UO_2 the numbers of ^{238}U nuclei are

$$N_{238} = 0.99275 \times N_A \frac{M}{A} = 0.99275 \times 6.022 \times 10^{23} \frac{1}{0.270} = 2.21 \times 10^{24},$$

where N_A is the Avogadro number and $A = (0.238 + 2 \times 0.016) = 0.270$ kg is the atomic mass of UO_2. From the preceding problem we know that 1 tonne of natural U_3O_8 contains 2.13×10^{27} uranium nuclei. Therefore to obtain 1 kg of UO_2 we need

$$(1000\,\text{kg}) \times 2.21 \times 10^{24}/2.13 \times 10^{27} = 1.038\,\text{kg of natural } U_3O_8.$$

Since the ore contains 0.1 % U_3O_8, 1.038 tonnes of ore are needed.
Let's assume that to enrich the UO_2 to 3 % we need to process an amount proportional to the increase in ^{235}U percentage. Then, $0.03/0.0072 = 4.17$ as much uranium with natural content of ^{235}U is needed, or

$$1.038 \times 4.17 = 4.328\,\text{tonnes of ore, containing a mass } M$$
$$= 4.328\,\text{kg of natural } U_3O_8.$$

From the preceding problem, we know that the activity of 1 kg of natural U_3O_8 is $R_{U\text{-nat}} = 1.52 \times 10^8$ Bq. Then, the 4.328 tonnes of uranium ore calculated above contain 0.1 % (= 4.328 kg) of U_3O_8, with a total activity of

$$R_{ore} = 4.328 \times 1.52 \times 10^8 \, \text{Bq} = 6.58 \times 10^8 \, \text{Bq}.$$

Then the specific activity of the uranium ore is

$$\frac{R_{ore}}{M} = \frac{6.58 \times 10^8}{4.328 \times 10^3} = 1.52 \times 10^5 \, \text{Bq/kg}.$$

Problems

6.1 Assume that in a reactor running on natural uranium (0.7 % ^{235}U and 99.3 % ^{238}U) the fission and capture processes are only induced by thermal neutrons. The fission and capture cross sections of ^{235}U are 580 b and 98 b, respectively and those for ^{238}U are 0 b and 2.7 b, respectively. Calculate the fraction of ^{238}U that would be transformed into ^{239}U (that is, finally into ^{239}Pu) when 1 % of ^{235}U has been used up. What quantity of ^{239}Pu would be produced in the case of a reactor containing initially 10 tonnes of natural uranium? (Assume the number of nuclei that can undergo capture or fission to be constant; this is a reasonable approximation if the number of transmuted nuclei is very small). [*Ans.*: about 0.004 %; about 10^{24} nuclei Pu]

6.2 Assume that in a reactor running on natural uranium (0.7 % ^{235}U and 99.3 % ^{238}U) only 1 % of the ^{235}U is fissioned. Calculate the mass of natural uranium that has the thermal energy equivalent of 1000 tonnes of coal, knowing that the energy equivalent of 1 kg of ^{235}U is about 3.4×10^6 kg of coal (see Problem 3.4).
[*Ans.*: 4200 kg]

6.3 Consider uranium enriched to 3 % of ^{235}U in a mixture with ^{238}U. Assume that in a reactor cycle, the percentage of ^{235}U decreases to 1 % (that is, 2/3 of the ^{235}U are fissioned).

(a) Knowing that the energy equivalent of 1 kg of ^{235}U is about 3.4×10^6 kg of coal, calculate what mass of uranium fuel has the thermal energy equivalent of 1000 tonnes of coal.

(b) If the fuel was coming from a *high-grade* uranium ore, that is, an ore containing 2 % uranium, what would be the amount of ore from which to extract a thermal energy equivalent to 1000 tonnes of coal?

[*Ans.*: (a) 14.7 kg of ^{235}U; (b) 735 kg uranium-ore]

References

1. IAEA Safety standards series No. GSG-1, Classification of Radioactive Waste, General Safety Guide, 2009
2. Radioactive waste in perspective, © OECD 2010, NEA No. 6350
3. http://www.world-nuclear.org/info/nuclear-fuel-cycle/nuclear-wastes/waste-management-over view/
4. http://newmdb.iaea.org/dashboard.aspx
5. http://www.world-nuclear.org/info/Nuclear-Fuel-Cycle/Nuclear-Wastes/Appendices/Radioactive-Waste-Management-Appendix-1–Treatment-and-Conditioning-of-Nuclear-Wastes/
6. https://www.oecd-nea.org/janis/janis-4.0/documentation/janis-4.0_manual.html
7. Advanced Nuclear Fuel Cycles and Radioactive Waste Management, OECD-NEA No. 5990, 2006
8. WNA: http://www.world-nuclear.org/info/Nuclear-Fuel-Cycle/Introduction/Physics-of-Nuclear-Energy/
9. European Commission, Radioactive Waste Management in the European Union (1998)
10. W. Moomaw, P. Burgherr, G. Heath, M. Lenzen, J. Nyboer, A. Verbruggen, Renewable energy sources and climate change mitigation, Special Report of the Intergovernmental Panel On Climate Change (IPCC), 2001, p. 19, and Annex II: Methodology p. 190. ISBN 978-92-9169-131-9. Online: https://www.ipcc.ch/pdf/special-reports/srren/SRREN_FD_SPM_final.pdf
11. The principles of radioactive waste management, Safety Series No. 111-F, IAEA 1995
12. P. Agostini et al., Nucleare da fissione, stato e prospettive, Ed. S. Monti. ENEA 2008, ISBN 88-8286-189-9
13. NEA-6885 Nuclear energy today, ISBN 978-92-64-99204-7, OECD 2012
14. http://www.wipp.energy.gov
15. A.P. Mesnhik, Scientific American, October 2005. http://www.world-nuclear.org/info/Nuclear-Fuel-Cycle/Power-Reactors/Nuclear-Power-Reactors/
16. H. Nifenecker, O. Meplan, S. David, Accelerator Driven Subcritical Reactors, Institute of Physics, Series in Fundamental and Applied Nuclear Physics (2003). ISBN 978-07-5030-743-7
17. A. Kochetkov et al., Current progress and future plans of the FREYA Project, in *Proceedings of the Second International Workshop on Technology and Components of Accelerator-driven Systems*, Nantes, France 21–23 May 2013
18. H.A. Abderrahim, MYRRHA a flexible and fast spectrum irradiation facility, in *Proceedings of the 11th International Topical Meeting on Nuclear Applications of Accelerators (AccApp 2013)*, Bruges, Belgium, 5–8 August 2013
19. H.A. Abderrahim, Future advanced nuclear systems and the role of MYRRHA as a waste transmutation R&D facility. in *Proceedings of the International Conference of Fast Reactors and related fuel cycles. Safe technology and sustainable scenarios*, vol. 1, ed. by S. Monti (IAEA, 2015, Paris, France) 4–7 March 2013, p. 69, STI/PUB/1665. ISBN 978-92-0-104114-2
20. Generation-IV International Forum, https://www.gen-4.org/

Erratum to: Energy from Nuclear Fission

Enzo De Sanctis, Stefano Monti and Marco Ripani

Erratum to:
E. De Sanctis et al., *Energy from Nuclear Fission*,
Undergraduate Lecture Notes in Physics,
DOI 10.1007/978-3-319-30651-3

The book was inadvertently published with incorrect figures 2.1 and 2.6 (caption) and inline equation under Problem 3.9. The figures and inline equation have been replaced in the original chapter.

The updated original online version for this book can be found at 10.1007/978-3-319-30651-3

E. De Sanctis (✉)
Laboratori Nazionali di Frascati, Istituto Nazionale di Fisica Nucleare, Frascati (Rome), Italy
e-mail: enzo.desanctis@lnf.infn.it

S. Monti
International Atomic Energy Agency, Vienna International Centre, Vienna, Austria
e-mail: S.Monti@iaea.org

M. Ripani
INFN Genova, Genova, Italy
e-mail: marco.ripani@ge.infn.it

© Springer International Publishing Switzerland 2016
E. De Sanctis et al., *Energy from Nuclear Fission*, Undergraduate Lecture
Notes in Physics, DOI 10.1007/978-3-319-30651-3_7

Glossary

Accelerator A device that uses electromagnetic fields to accelerate charged particles to high energies.

Activity Term used to characterise the number of nuclei which disintegrate in a radioactive substance per unit time. It is usually measured in becquerels (Bq), 1 Bq is equal to 1 disintegration per second.

Advanced Gas-cooled Reactor (AGR) Reactor that uses enriched uranium as fuel, graphite as moderator, and CO_2 gas as coolant.

Accelerator Driven System (ADS) A subcritical reactor (that is, a reactor that cannot sustain a chain reaction) coupled to a high-energy accelerator. The latter is used to supply, through nuclear reactions, enough external neutrons to keep the chain reaction going. Accelerator Driven Systems are widely considered as promising devices for the transmutation of nuclear waste, as well as useful schemes for thorium-based energy production.

ALARA (As Low As Reasonably Achievable) Making every reasonable effort to minimize exposure to ionising radiation, so as to keep it below regulatory or legal dose limits, but taking into account the economic and social benefits of the use of radiation.

Alpha Decay The radioactive process in which an alpha particle (that is, a helium nucleus) is emitted by an unstable nucleus. During alpha decay the atomic number Z is reduced by two units and the mass number A by four units.

Alpha particle (symbol α) Positively charged particle emitted from the nucleus of some radioactive elements. It is identical to a helium nucleus, which has mass number 4 and consists of two protons and two neutrons. It has low penetrating power and a short range (a few centimetres in air). The most energetic alpha particles generally fail to penetrate the dead layers of cells covering the skin, and can be easily stopped by a sheet of paper.

Atom Smallest particle of an element that cannot be divided any further by chemical means. It consists of a central core (or nucleus) containing protons and

© Springer International Publishing Switzerland 2016
E. De Sanctis et al., *Energy from Nuclear Fission*, Undergraduate Lecture
Notes in Physics, DOI 10.1007/978-3-319-30651-3

neutrons, with positive electric charge, surrounded by distributions of negative electric charges, the electrons.

Atomic number (Z) The number of positively charged protons in the nucleus of an atom. It is equal to the number of electrons in that atom. It is usually denoted by Z.

Background (radiation) Natural radiation that is always present in the environment. It includes cosmic radiation, which comes from the space, terrestrial radiation, which comes from the Earth, and internal radiation, which exists in all living things.

Barn (b) Unit of area. In nuclear physics, it is used as unit of measure of the cross section of a nuclear reaction, which is related to the probability of occurrence of that reaction. 1 barn is equal to 10^{-28} m^2.

Base load power plants Power plants for electricity supply that, due to their operational and economic properties, are used to cover the base load, i.e. the minimum level of demand on an electrical grid over 24 h. Generally, they are hydroelectric, coal, oil, gas-fired and nuclear power plants.

Becquerel (Bq) International system (SI) unit of measure of radioactivity. 1 Bq represents a rate of radioactive decay equal to 1 disintegration of a nucleus per second. It is a very small unit, therefore in practice, megabecquerel (MBq), gigabecquerel (GBq) or terabecquerel (TBq) are the more common units.

Beta decay Radioactive process in which a beta particle is emitted. During beta decay the mass number A of the newly created nuclide is equal to that of the parent nuclide, whereas the atomic number Z changes by one unit. Specifically, the atomic number during beta decay emitting a positron (or antielectron) (β^+ decay) becomes smaller by one unit and during beta decay emitting an electron (β^- decay) greater by one unit.

Beta particle Charged particle (with a mass equal to about 1/1836 that of a proton) emitted during the β decay of a radioactive nucleus. Specifically, the negatively charged beta particle is identical to an electron, while the positively charged beta particle is called a positron (antielectron).

Beyond design-basis accident (BDBA) A nuclear installation is designed and built to withstand certain design-basis accidents (see corresponding definition). Very unlikely events not considered in the design-basis and resulting in accidental scenarios are called beyond design-basis accidents.

Binding nuclear energy Minimum energy required to separate the nucleus of an atom into its component neutrons and protons.

Blanket Specific component of fast reactors containing fertile material for breeding of fissile material.

Boiling Water Reactor (BWR) Nuclear reactor that uses water as coolant and moderator, where water is boiling in the core. The steam produced is then directly used to drive turbines, which activate generators to produce electrical power, with no secondary fluid loops.

Boron (symbol B) Metalloid chemical element with atomic number 5. In nuclear reactors, it is used as a neutron absorber.

Breeding Conversion of non-fissionable material into fissionable material, e.g. uranium-238 into plutonium-239 or thorium-232 into uranium-233.

Breeding ratio Ratio of bred fissile material to burnt fissile material after the irradiation of a fuel mixture of fissile and fertile material in a reactor.

Breeder reactor Reactor that produces more nuclear fuel than it consumes. Typically, such reactors have fertile material placed in and around the reactor core in order to use neutrons produced during fission to transmute the fertile material into fissile material. For example, uranium-238 can be placed around a fast reactor and it will undergo transmutation producing plutonium-239, which can then be recycled and used as fuel in the reactor.

Burnup Total energy released in the fission of a given amount of nuclear fuel. It is usually measured in megawatt-days (abbreviated as MWd). The fission energy release per unit mass of the fuel is termed *specific burnup* of the fuel and is usually expressed in megawatt-days per kilogram of the heavy metal originally contained in the fuel, abbreviated as MWd/kgHM, where HM stands for heavy metals. Often used is the practical unit megawatt-days per tonne (MWd/tHM).

Cadmium (symbol Cd) Soft, bluish-white metal chemical element with atomic number 48. In nuclear reactors, it is used as a neutron absorber.

CANDU reactor (CANadian Deuterium Uranium reactor) Reactor using heavy-water, i.e. deuterium oxide, as coolant and moderator. The use of heavy water results in less neutron absorption and permits the use of natural uranium as the reactor fuel, thereby eliminating the need for enrichment of the uranium.

Capacity factor It is the ratio of actual output of a power plant over a period of time, to its potential output if it were possible for it to operate at full nominal capacity continuosly over the same period of time. To calculate it, take the total amount of energy the plant produced during a period of time and divide by the amount of energy the plant would have produced at full capacity.

Cask Heavily shielded cylindrical container (often made of lead, concrete, or steel) used for the dry storage or shipment (or both) of radioactive materials such as spent nuclear fuel or other high-level radioactive waste.

Centrifuge Device used to separate the uranium isotopes (as gases) based on their slight difference in mass. It is used in the enrichment process to prepare uranium for the fabrication of nuclear fuel. This process uses a large number of inter-connected centrifuge machines.

Chain reaction A reaction that initiates and mantains its own repetition. In a nuclear reactor or critical assembly, a fissile nucleus absorbs a neutron and fissions spontaneously, releasing additional neutrons. These, in turn, can be absorbed by other fissile nuclei, releasing still more neutrons. A fission chain reaction is self-sustaining if, on average, for every fission that occurs, exactly one emitted neutron will produce another fission. If more than one emitted neutron per fission will produce another fission, the number of fissions will rapidly grow. If less than one emitted neutron per fission will produce another fission, the number of fissions will rapidly die away.

Chernobyl Location, 130 km north-west of Kiev, Ukraine, where on April 26, 1986 the most serious accident to date in the peaceful use of nuclear energy occurred.

Cladding Tightly bound (usually metallic) casing directly surrounding the nuclear fuel, which protects it against a chemically active ambience (e.g. cooling water) and prevents the escape of fission products into the coolant.

Closed fuel cycle Fuel cycle that recovers fissile material from spent fuel, and uses it to fabricate new fuel to be used again in a reactor.

Comprehensive Nuclear-Test-Ban Treaty (CTBT) Multilateral treaty by which States agree to ban all nuclear explosions in all environments, for military or civilian purposes. CTBT was adopted by the United Nations General Assembly on September 10, 1996, but has not entered into force as eight specific states have not yet ratified the treaty.

Containment building Steel-reinforced concrete structure that houses a nuclear reactor and its pressurizer, reactor coolant pumps, steam generator, and other equipment or piping that might otherwise release fission products to the atmosphere in the event of an accident.

Contamination Deposition of, or presence of radioactive substances on surfaces or within solids, liquids or gases (including the living organisms and the human body), where their presence is unintended or undesirable.

Control rod Rod, plate, or tube containing a material (such as boron, silver, indium, cadmium and hafnium) which absorbs neutrons. Control rods are used to reduce the number of neutrons going around in the reactor core. They can be used to tune the power level and power spatial distribution in the reactor during operation or to completely stop the fission chain reaction, thereby shutting down the reactor, when necessary.

Conversion Chemical process used to turn solid uranium oxide received from a uranium mill into volatile uranium hexafluoride, which is a gas at certain temperatures and pressures, and therefore suitable for the enrichment process.

Converter reactor Nuclear reactor which generates fissile material, but less than the quantity it burns.

Coolant Substance circulated through the core of a nuclear reactor to remove or transfer the thermal energy. The most commonly used coolant is water. Other coolants include heavy-water, liquid sodium, helium gas, carbon dioxide, liquid lead or a liquid lead-bismuth eutectic mixture. A coolant can also be a moderator; water serves this dual function in most reactors.

Core Central portion of a nuclear reactor, which contains the fuel assemblies, the moderator, the control rods, the coolant and the support structures. The reactor core is where fissions occur.

Cosmic radiation Radiation that originates directly or indirectly from extra-terrestrial sources. Cosmic radiation is part of natural radiation. Differences in elevation, atmospheric conditions, and the Earth's magnetic field can change the amount (or dose) of cosmic radiation that we receive.

Criticality Normal operating condition of a reactor, in which nuclear fuel sustains a steady-state fission chain reaction. A reactor achieves criticality (and is said to be critical) when neutrons lost by leakage or absorption leave exactly one emitted neutron per fission available to produce another fission, such that the number of neutrons produced in fission remains constant and so does the power.

Critical mass The smallest mass of fissionable material that will support a self-sustaining chain reaction for a given set of conditions, e.g. shape of the fissionable material, amount and type of moderator or reflector.

Cross section In atomic, nuclear and particle physics, fictitious area related to the probability of occurrence of a reaction. It depends on the nature of the colliding particles and the forces between them.

Curie (Ci) Historical unit of measure of radioactivity. 1 Ci is equal to 37 billion (3.7×10^{10}) disintegrations per second, so 1 Ci also equals 3.7×10^{10} becquerels (Bq).

Dating, radioactive Method of measuring the age of an object by determining the ratio of various radionuclides to stable nuclides contained therein.

Daughter product Nuclide formed by the radioactive decay of some other nuclide (parent nucleus). In the case of uranium-238, for example, there are 14 successive daughter products, ending with the stable isotope lead-206.

Decay (radioactive) Spontaneous conversion of a nuclide into another nuclide or into another energy state of the same nuclide. Every decay process has a certain half-life.

Decay constant Characteristic, well-defined probability of decay per unit time of a radioactive nucleus. It is an intrinsic property of the nucleus itself, which is independent of its physical and chemical conditions. It is usually denoted by the Greek letter λ.

Decay heat Thermal energy produced by the decay of radioactive fission products and actinides from reactor operation.

Decommissioning Process where a nuclear power plant (or other facility where nuclear materials are handled) is safely and permanently shut down (and possibly dismantled) to retire it from service after its useful life has ended. Depending on specific policies and decisions, decommissioning can involve several stages: closeout, decontamination and dismantling, and demolition and site clearance.

Defence-in-depth Approach to design and operation of nuclear facilities that prevents and mitigates the consequences of accidents. It is based on the use of multiple independent and redundant layers of protection to prevent and mitigate the release of radiation or hazardous materials. It includes the use of physical and administrative controls, physical barriers, redundant safety functions and emergency response measures.

Depleted uranium Uranium with percentage of ^{235}U lower than the 0.72 % contained in natural uranium. Depleted uranium is produced as a by-product of the enrichment process and can also be found in mine tailings or residues.

Design basis accident (DBA) Range of conditions and events (e.g. rupture of piping, coolant pump failure) taken explicitly into account in the design of a nuclear facility, such that the facility can withstand without fatal consequences for the systems, structures, and components necessary to ensure safety. Design Basis Accidents typically include external occurrences like earthquakes, floods, severe weather events, etc., as well as internal ones like fires, uncontrolled reactivity (and power) surges, etc.

Deterministic effects Effects that will certainly occur (e.g. measurable changes in blood parameters), should a radiation exposure exceed the threshold for that effect. The magnitude of the effect is proportional to the exposure above the threshold.

Deterministic safety approach Method of assessing the safety of a nuclear power plant using a defined set of initiating events, "design basis events". The latter are chosen to encompass a range of realistic possible initiating events that could challenge the safety of the plant. Examples include loss-of-coolant accidents, control rod ejection (for a PWR), control rod drop (for a BWR) and steam line break. Engineering analysis is used to predict the response of the plant and its safety systems to the design basis events and to verify that this response remains within prescribed regulatory limits.

Deuterium (symbol $^{2H \ or \ D}$) Stable isotope of hydrogen with one proton and one neutron in its nucleus compared with the one proton in the nucleus of ordinary hydrogen. It is therefore also called *heavy hydrogen*. Deuterium has a natural abundance in Earth's oceans of about one atom in 6420 of hydrogen.

Deuteron (symbol *d*) Nucleus of deuterium. It contains one proton and one neutron.

Dose, absorbed Amount of energy absorbed by an object or person per unit mass. It reflects the amount of energy that ionising radiation sources deposit in materials through which they pass, and is measured in gray (Gy). 1 Gy = 1 J/kg.

Dose, equivalent Measure of the biological damage to living tissue because of radiation exposure. It accounts for the different biological effectiveness of different radiation. It is the product of the organ absorbed dose and the radiation weighting factor. It is measured in sievert (Sv).

Dose, effective Sum of the average organ equivalent doses in the individual organs and tissues of the body, due to external or internal radiation exposure, multiplied by the tissue weighting factors. It is the suitable quantity to indicate a uniform dose value in case of different exposure of various body parts, in order to evaluate the risk of late radiation injuries. It is measured in sievert (Sv).

Dosimetry Measuring procedure to determine the dose equivalent generated by ionizing radiation in matter.

Doubling time Operating time during which a breeding reactor will produce enough additional fissile material to operate another reactor for a same time.

Dry storage Following an initial cooling period in a water pool, spent fuel can be loaded into large, shielded casks in which natural air circulation maintains it at the required temperatures.

Electron (symbol *e*) Elementary particle with a negative charge and a mass equal to 1/1836 that of a proton. Electrons surround the positively charged nucleus of an atom, and determine its chemical properties.

Electron capture Process that unstable atoms can use to become more stable. During electron capture, an atomic electron interacts with the nucleus where it combines with a proton, forming a neutron and a neutrino. The neutrino is ejected from the nucleus. Since an atom loses a proton during electron capture, it changes from one element to another. For example, after undergoing electron capture, an atom of carbon (with 6 protons) becomes an atom of boron (with 5 protons). Although the numbers of protons and neutrons in the nucleus change during electron capture, the total number of nucleons (protons + neutrons) remains the same.

Electronvolt (eV) Commonly used unit of energy in atomic, nuclear and particle physics. It is the kinetic energy that an electron acquires when it is accelerated by a potential difference of 1 V (1 eV = 1.602×10^{-19} J). Widely used are the multiples: kiloelectronvolt (keV), equal to 1000 eV; megaelectronvolt (MeV), equal to 10^6 eV; and gigaelectronvolt (GeV), equal to 10^9 eV.

Element Chemical base material that cannot be chemically converted into simpler substances. Examples: carbon, oxygen, iron, mercury, lead, uranium.

Enriched uranium Uranium in which the isotopic concentration of fissile isotope uranium-235 has been increased above the naturally occurring level of 0.72 %.

Enrichment (uranium) Physical process by which the isotopic concentration of fissile isotope uranium-235 is increased above the level found in natural uranium. Two enrichment processes are commercially used, gaseous diffusion and gas centrifugation.

Evolutionary Power Reactor (EPR) A new Generation III + Pressurised Water Reactor (PWR). It is also called European Pressurised Reactor.

Equilibrium, transient A condition that occurs during a radioactive decay chain, when the half-life of the starting nuclide is greater than the half-lives of the decay products. After a period much longer than the longest half-life of the decay products, the ratios between the activities of the members of the decay chain become constant in time. Moreover, the activity of a daughter becomes greater than that of the parent.

Equilibrium, secular In a series decay, it is the situation in which the quantity of a radioactive isotope remains constant because its production rate (e.g., due to decay of a parent isotope) is equal to its decay rate. For instance, it occurs when a long-lived isotope generates a family of subsequent, shorter-lived daughters.

Excited state Configuration of an atom or a nucleus with a higher energy than that of its lowest-energy state (*ground state*). The excess energy is generally emitted as photons (gamma quanta).

Fallout Radioactive material which falls back to the ground after the release into the atmosphere (e.g. by nuclear weapon tests, accidents).

Fast neutron Neutron with a kinetic energy above 10 keV and typically about 2 MeV. Fast neutrons can cause fission in fissile materials but the probabilities (cross sections) are lower than that for thermal neutrons.

Fast reactor Reactor designed for criticality and operation based on fast neutrons. A fast reactor has no moderator since deceleration of the fast fission neutrons generated during the fission should be avoided. It must use a nuclear fuel containing a higher percentage of fissile nuclei with respect to that of thermal reactors.

Fertile material Material which is not itself fissile, that can be converted into a fissile material through the capture of a neutron, possibly followed by radioactive decay. Important examples are uranium-238 and thorium-232. When they capture neutrons, they are converted after a few decays into fissile plutonium-239 and uranium-233, respectively.

Fissile material Material that is capable of undergoing fission after the capture of a low-energy thermal (slow) neutron. In practice, the most important fissile materials are uranium-233, uranium-235 and plutonium-239.

Fission Process through which the nucleus of an atom splits into two or more fragments accompanied by the release of neutrons, gamma radiation, and significant amounts of energy. Fission may be spontaneous for a heavy nucleus, but it is usually caused by the nucleus of an atom becoming unstable after absorbing a neutron.

Fission products Nuclides formed by fission of heavy elements, plus nuclides formed by subsequent radioactive decay of fission fragments. Fission products may be stable or unstable. They and their decay products form a significant component of nuclear waste.

Fission yield Percentage of appearance of a fission product in nuclear fission. Fission products with mass numbers A between 88 and 103 and between 132 and 147 have particularly high fission yields.

Fissionable material Nuclide capable of undergoing fission after capturing either high-energy (fast) neutrons or low-energy thermal (slow) neutrons, however with cross sections that are much smaller than the fission cross sections of fissile nuclei for thermal neutrons. An example of a fissionable material is uranium-238.

Fuel (nuclear) Component of a reactor, usually in the solid state, consisting of fissile elements (e.g. ^{233}U, ^{235}U, ^{239}Pu, ^{241}Pu), where the fission reactions and the transformation of fission energy into thermal energy occurs.

Fuel assembly (fuel bundle) Structured group of fuel rods, that are long, slender metal tubes containing the nuclear reactor fuel in form of pellets with a certain content of fissile material. Different reactor designs feature a different number of fuel assemblies in the core, a different number of fuel rods per assembly and a different size of the fuel rods.

Fuel cycle Series of steps involved in creating, using and disposing of fuel for nuclear power reactors. It includes mining and milling of uranium, conversion, enrichment, fabrication of fuel elements, use in a reactor, possibly reprocessing and finally, waste disposal.

Fuel fabrication Manufacture, processing, and assembly of fuel elements for reactors.

Fuel reprocessing (see reprocessing).

Fuel rod Hollow, slender metal tube containing the nuclear reactor fuel in form of pellets with a certain content of fissile material. The diameter and length of the rods and their number in an assembly are different in different reactors (see Fuel assemblies).

Fukushima-Daiichi Nuclear power plant located in the Futaba District of Fukushima Prefecture, on the Honshu Island (the main part of Japan), where on March 11, 2011 a serious nuclear accident occurred following a great earthquake (magnitude 9 on the Richter's scale) and a subsequent gigantic anomalous wave (a so-called tsunami).

Fusion Nuclear reaction where light nuclei combine to form more massive nuclei with release of energy. This process takes place continuously in the universe. In the core of the Sun, at a temperature of 15 million degrees celsius, hydrogen is converted to helium, providing the energy that sustains life on Earth.

Gas-cooled reactor Nuclear reactor cooled with gas (helium, carbon dioxide).

Gamma rays High-energy, short-wavelength, electromagnetic radiation emitted from the nucleus of an atom. Gamma radiation frequently accompanies emissions of alpha particles and beta particles, and always accompanies fission. Gamma rays are similar to X-rays, but are very penetrative and may be best stopped or shielded by materials of high density and high atomic number, such as lead or depleted uranium.

Geological repository Excavated, underground storage facility designed, constructed, and operated for safe and secure permanent disposal of high-level radioactive waste. A geological repository uses an engineered barrier system and a portion of the site's natural geology, hydrology, and geochemical systems to isolate the radioactivity of the waste.

Generation-I (Reactors) Represents the first batch of nuclear power reactors. In many countries, they were experimental reactors derived from smaller naval propulsion cores. In a few countries, they represented the first civilian nuclear power plants.

Generation-II (Reactors) It is a group of the second batch of nuclear power plants built in the world. These reactors were explicitly built for electricity generation. Most of the nuclear power plants in operation today belong to Generation-II.

Generation-III (Reactors) Represents a set of standardised light-water nuclear reactor designs with increased safety and economic efficiency. These reactors are expected to ensure the core integrity in case of a serious external event like an aircraft fall or an earthquake.

Generation-IV (Reactors) Innovative reactor designs that could be commercially deployed starting from 2040. They should show increased sustainability, competitive economics, high level of safety, increased proliferation resistance, and the ability to cogenerate high-grade heat for use in industrial processes.

GIF (Generation IV International Forum) International co-operative endeavour set up to carry out the research and development needed to establish the feasibility and performance capabilities of the next generation nuclear energy systems (Generation IV reactors). The GIF has thirteen Members.

GeV (gigaelectronvolt) Unit of energy equivalent to one billion electronvolts; $1 \text{ GeV} = 10^9 \text{ eV}$.

GJ (gigajoule) Unit of energy equivalent to one billion joules; $1 \text{ GJ} = 10^9 \text{ J}$.

Graphite Form of carbon, similar to that used in pencils, used as a moderator in some nuclear reactors.

Gray It is the international system (SI) unit of absorbed radiation dose. It is equal to one joule per kilogram of absorbing medium; 1 Gy = 1 J/kg.

GW **(gigawatt)** Unit of power equivalent to one billion watts. 1 GW = 1000 MW = 10^6 kW = 10^9 W.

GWd **(Gigawatt-day)** Unit of energy corresponding to a power of 1 GW (=10^9 J/s) extended over one day (=86,400 s); 1 GWd = 8.64 · 10^{13} J.

GWh **(Gigawatthour)** Unit of energy equivalent to one billion watthours; 1 GWh = 3.6 × 10^{12} J.

Hafnium (symbol Hf) Chemical, metal element with atomic number 72. In nuclear reactors, hafnium is preferably used as a neutron absorber to avoid criticality incidents.

Half-life It is the time required for one-half of the radioactive isotopes in a sample to decay to (or disintegrate into) other nuclear isotopes.

Heavy water (D_2O) It is water containing significantly more than the natural proportions (one in 6420) of deuterium atoms to ordinary hydrogen atoms. Heavy water is used as a coolant and moderator in pressurised heavy water reactors (PHWRs) because its properties allow natural uranium to be used as fuel.

Heavy-water reactor Reactor cooled and/or moderated with heavy water (D_2O).

Highly-enriched uranium (HEU) Uranium enriched to at least 20 % uranium-235.

High-level waste (HLW) Highly radioactive material produced by the reactions that occur inside nuclear reactors or as by-products of fuel reprocessing. In general, HLW contains long-lived radionuclides with high activity, which may also produce heat. Geological disposal is foreseen for this type of waste.

IAEA (International Atomic Energy Agency) Autonomous, intergovernmental organisation set up in 1957 by the United Nations with the mission to promote the safe, secure and peaceful use of nuclear technologies and to inhibit its use for any military purpose. As of the end of September 2015, it had 165 member states. It is based in Wien, Austria (https://www.iaea.org/).

IEA (International Energy Agency) Autonomous, international organisation (founded in 1974, http://www.iea.org/), which works to ensure reliable, affordable and clean energy for its 29 member countries and beyond. It has four main areas of focus: energy security, economic development, environmental awareness and engagement worldwide.

INES (International Nuclear Event Scale) Scale with seven levels proposed by the IAEA and NEA to evaluate the events occurring in nuclear installations according to international uniform criteria, in particular with regard to the aspect of hazards for the population.

Ingestion Intake of—radioactive-substances through food and drinking water.

Inhalation Intake of—radioactive-substances with breathing.

Intermediate-level waste (ILW) Radioactive waste is normally classified into a small number of categories to facilitate regulation of handling, storage and disposal based on the concentration of radioactive material it contains and the time for which it remains radioactive. The definitions of categories differ from country to country. However, in general, ILW needs specific shielding during handling and, depending on the specific content of long-lived radionuclides, it may need geological disposal or it may be suitable for surface or near-surface disposal.

Internal conversion Process that unstable atoms can use to become more stable. During internal conversion, an excited nucleus de-excites by knocking out one of the electrons in the atom. The electron is emitted with a well-defined energy. During internal conversion, the atomic number does not change.

Ion Atom with one or more electrons less or one or more electrons more, causing it to have an electrical charge, and therefore, to be chemically active.

Ionising radiation Any form of radiation, either particles or electromagnetic waves that, when crossing matter, can deposit enough energy to produce ions by breaking molecular bonds and displace (or remove) electrons from atoms or molecules.

Irradiation Any form of exposure of matter (including living tissues) to ionizing radiation. Also fuel that spent some time in a reactor core is called irradiated fuel.

Isobars Nuclei with the same number of nucleons are isobars. For example, nitrogen-17 (^{17}N), oxygen-17 (^{17}O) and fluorine-17 (^{17}F) are isobars. All three nuclei have 17 nucleons, the nitrogen nucleus however has 7 protons and 10 neutrons, the oxygen nucleus has 8 protons and 8 neutrons, and the fluorine nucleus has 9 protons and 8 neutrons.

Isotope Different atomic configurations of a given element that have the same number of protons but different number of neutrons in their nuclei are isotopes. They have the same or very similar chemical properties and distinct physical properties. For example, deuteron and tritium are isotopes of hydrogen, having in their nuclei respectively one and two neutrons in addition to a proton. Uranium-235 (^{235}U) and uranium-238 (^{238}U) are both isotopes of uranium with ^{235}U having 143 neutrons and ^{238}U, 146. Among their distinct physical properties, some isotopes are radioactive, while others are not. For example, carbon-12 (^{12}C) and carbon-13 (^{13}C) are stable, but carbon-14 (^{14}C) is unstable and radioactive.

keV (kiloelectronvolt) Unit of energy equivalent to one thousand electronvolt; 1 keV = 1000 eV.

kJ (kilojoule) Unit of energy equivalent to one thousand joule; 1 kJ = 1000 J.

kW (kilowatt) Unit of power equivalent to one thousand watt; 1 kW = 1000 W.

kWh (kilowatthour) Unit of energy equivalent to one thousand watthour; 1 kWh = 3.6×10^6 J.

Latent cancer fatality (LCF) Death resulting from a cancer that developed after a latent period following exposure to radiation.

Lethal dose Ionizing radiation dose leading to the death of the irradiated individual due to acute radiation injuries. The average lethal dose (LD_{50}) is the dose at which on average half of the individuals with similar irradiation quantities die.

Light-water (H_2O) Ordinary water as opposed to heavy water.

Light water reactor (LWR) Nuclear reactor type that is cooled and/or moderated by ordinary water, as opposed to heavy water.

Loading Factor (see capacity factor).

Loss of coolant accident (LOCA) Postulated accident that results in a loss of reactor coolant not mitigated by safety systems.

Long-term operation (LTO) Term generally used to describe the operation of nuclear power plants beyond their original design lifetime. It involves specific plant life management issues such as safety upgrades, inspection of critical equipment (for instance the pressure vessel), replacement of large components (for example steam generators or turbine modules) and possibly power uprates.

Low-enriched uranium (LEU) Uranium in which the isotopic concentration of uranium-235 has been increased above naturally occurring levels while remaining less than 20 %. Typically, nuclear power reactors use low-enriched uranium with 3–5 % uranium-235.

Low-level waste (LLW) In general, LLW is a type of waste that does not need significant shielding for handling and, because of the absence of long-lived radionuclides, is suitable for surface or near-surface disposal. About 90 % of the radioactive waste volume produced in the world each year is LLW. Some examples include radioactively contaminated protective shoe covers and clothing; cleaning rags, mops, filters, and reactor water treatment residues; equipment and tools; medical tubes, swabs, and hypodermic syringes; and carcasses and tissues from laboratory animals.

Mass number Number of nucleons (neutrons and protons) in the nucleus of an atom. For example, the mass number of U-238 is 238 (92 protons and 146 neutrons). It is usually denoted by A.

Mean-life Period of time during which the number of radionuclide nuclei reduces to $1/e$ (where $e = 2.718$ is the base of natural logarithms). It is equal to the reciprocal of the decay constant λ.

MeV (megaelectronvolt) Unit of energy equivalent to one million of electron-volts; 1 MeV $= 10^6$ eV.

Milling Process through which mined uranium ore is chemically treated to extract and purify the uranium. It also reduces the volume of material to be transported and handled in fuel manufacture. Reflecting its colour and consistency, the solid product (U_3O_8) of milling is known as *yellow cake*.

Mill tailings The remnant of a metal-bearing ore consisting of finely ground rock and process liquid after some or all of the metal, such as uranium, has been extracted.

Mixed-oxide fuel (MOX) A fuel for nuclear power plants that consists of a mixture of depleted uranium oxide and plutonium oxide.

Moderator Material used in a thermal reactor to slow down high-velocity neutrons to the thermal energy range so as to increase their efficiency in causing fission. The moderator must be a light material that will allow the neutrons to slow down efficiently without having a high probability to be absorbed. Usually, ordinary water is used; an alternative in use is heavy-water or graphite.

MW (Megawatt) Unit of power equivalent to one million of watt; 1 MW $= 10^6$ W.

MWd (megawatt day) Unit of energy corresponding to a power of 1 MW ($=10^6$ J/s) extended over one day ($=86{,}400$ s). 1 MWd $= 8.64 \times 10^{10}$ J.

MWh (Megawatthour) Unit of energy equivalent to one million watthours; 1 MWh $= 10^6$ Wh $= 3.6 \times 10^9$ J.

MW(e) (Megawatt electric) Electric output power in megawatt of a plant. It is equal to the thermal overall power multiplied by the thermal efficiency of the plant. The power plant efficiency of light water reactors amounts to 30–33 % compared to up to 40 % for modern coal-, oil- or gas-fired power plants.

MW(th) (Megawatt thermal) Overall thermal power output of a nuclear reactor in megawatt. Typically, the thermal power of a nuclear reactor is about three times its electrical power.

MWd/t (Megawatt day per tonne) Unit for the thermal energy output for one tonne of nuclear fuel during the service time in the reactor.

Nanometre (nm) Unit of length equal to one billionth of metre; 1 nm $= 10^{-9}$ m.

Natural uranium Uranium with the isotope composition occurring in nature. It is a mixture of uranium-238 (99.275 %), uranium-235 (0.720 %), and uranium-234 (0.005 %).

NEA (Nuclear Energy Agency) Specialised agency within the Organisation for Economic Co-operation and Development (OECD) based in Paris, France (http://www.oecd-nea.org/). Its mission is to assist its member countries in maintaining and further developing the scientific, technological and legal bases required for a safe, environmentally friendly and economical use of nuclear energy for peaceful purposes.

Neutrino (symbolν) Electrically neutral elementary particle with almost zero mass, that interacts with matter through the nuclear weak interaction.

Neutron (symbol$n$) Elementary particle with no electric charge and a mass (1.67493×10^{-27} kg) slightly greater than a proton, found in the nucleus of all atoms except hydrogen-1 (^{1}H). The free neutron is unstable and decays with a half-life of 10.17 minutes (mean-life 880.3 s).

Neutron capture Reaction that occurs when a nucleus captures a neutron.

Neutron, delayed Neutron not generated directly during the nuclear fission, but delayed because it originates from the decay of neutron-rich fission fragments, or their daughters, which have a half-life of the order of seconds. Less than 1 % of the neutrons involved in nuclear fission are delayed.

Neutron emission One process that unstable atoms can use to become more stable. During neutron emission, a neutron is ejected from an atomic nucleus. Since the number of protons within an atom does not change during neutron emission, the atom does not change from one element to another. However, it becomes a different isotope of the parent element. For example, after undergoing neutron emission, an atom of beryllium-13 (with 9 neutrons) becomes an atom of beryllium-12 (with 8 neutrons).

Neutron flux Measure of the intensity of neutron radiation, determined by the neutron flow rate. The neutron flux value is calculated as the neutron density (number of neutrons per unit volume n) multiplied by neutron velocity (v). Consequently, neutron flux (nv) is typically measured in neutrons/(cm^2 s).

Neutron, prompt Neutron emitted immediately during the nuclear fission (within about 10^{-14} s); in contrast to delayed neutrons which are emitted seconds to minutes after the fission by fission products. Prompt neutrons make up more than 99 % of the neutrons.

Neutron, slow Neutron whose kinetic energy falls below a certain value—frequently 10 eV is selected.

Neutron, thermal Neutron in thermal equilibrium with the medium in which it moves. It has a mean energy $\sim kT$ (where $k = 1.38 \times 10^{-23}$ JK^{-1} is the Boltzmann's constant) given by the temperature T of the medium; at ordinary temperature, $T = 300$ °C, $kT \approx 0.025$ eV. Thermal neutrons have the greatest probability of causing fission in uranium-235 and plutonium-239.

Non-ionising radiation Any form of radiation, either particles or electromagnetic waves, with insufficient energy to ionise atoms. Examples of non-ionising radiation include radio waves, light and microwaves.

Non-Proliferation Treaty (NPT) The Treaty on the Non-Proliferation of Nuclear Weapons, commonly known as the Non-Proliferation Treaty, is an international treaty whose objective is to prevent the spread of nuclear weapons and weapons technology, to promote cooperation in the peaceful uses of nuclear energy, and to further the goal of achieving nuclear disarmament and general and complete disarmament.

Nuclear medicine Application of radioactive substances and radiation sources in medicine for diagnostic or therapeutic purposes.

Nuclear poison Isotope with a large neutron absorption cross section. If present in the reactor core, nuclear poisons absorb neutrons, subtracting them from the fission chain, thereby decreasing the reactor multiplication coefficient, or reactivity. Certain fission products generated during fission, such as xenon-135 and samarium-149, are poisons, so that at high concentration they can even stop the reactor from producing energy.

Nuclear reactor Device that uses the nuclear fission process to produce energy. Though there are many types of nuclear reactors, they all incorporate certain essential features, including the use of fissile material as fuel, a moderator to increase the probability of fission (unless the reactor uses fast neutrons), a coolant for thermal energy removal, and control rods. Other common features include a reflector to bounce back escaping neutrons, shielding to protect personnel from radiation exposure, instrumentation to measure and control the reactor, and devices to protect the reactor.

Nucleon Common name for proton and neutron, constituent particles of the atomic nucleus.

Nucleus Small, central, positively charged region of an atom. Except for the nucleus of ordinary hydrogen, which has only a proton, all atomic nuclei contain both protons and neutrons. The number of protons determines the total positive charge or atomic number Z. This number is the same for all the atomic nuclei of a given chemical element. The number of protons plus neutrons is called the mass number, A.

Organization for Economic Cooperation and Development (OECD) Intergovernmental organization (based in Paris, France, http://www.oecd.org/) which provides a forum for discussion and cooperation among the governments of industrialized countries committed to democracy and the market economy. Its primary goal is to support sustainable economic growth, boost employment, raise living standards, maintain financial stability, assist other countries' economic development, and contribute to growth in world trade. In addition, the OECD is a reliable source of comparable statistics and economic and social data.

The OECD also monitors trends, analyses and forecasts economic developments, and researches social changes and evolving patterns in trade, environment, agriculture, technology, taxation, and other areas.

Once-through fuel cycle Nuclear fuel cycle in which spent fuel is not recycled. Once removed from the reactor, the spent fuel is conditioned and stored until a disposal repository becomes available.

Partitioning and transmutation (P&T) Partitioning is the separation of long-lived radioactive nuclides (minor actinides and fission products) from spent fuel. Transmutation is their transformation by means of nuclear reactions into short-lived or stable nuclides. P&T is proposed as a methodology to reduce the radiotoxicity of nuclear waste and reduce the amount destined to geological disposal.

Pebble bed reactor Gas-cooled high-temperature reactor with a core of spherical fuel pebbles and graphite as moderator.

Pellet, fuel Thimble-sized ceramic cylinder consisting of enriched uranium or MOX fuel, filled into the fuel cladding tubes.

PET (Positron Emission Tomography) Functional imaging technique used in nuclear medicine to observe metabolic processes in the body. PET detects pairs of gamma rays emitted by electron-positron annihilation caused by a positron-emitting radionuclide (tracer), which is introduced into the body on a biologically active molecule. Three-dimensional images of tracer concentration within the body are then constructed by computer analysis.

Photon (symbolγ) Energy quantum of electromagnetic radiation. It has rest mass equal to zero and no electrical charge.

Pitchblende (see yellow cake).

Plutonium (symbol Pu) Heavy, radioactive, manmade metallic element with atomic number 94. Its most important isotope is fissile plutonium-239, which is produced by neutron irradiation of uranium-238. It exists only in trace amounts in nature.

Positron (symbol e^+) Elementary particle with the mass of an electron, but positively charged. It is the antiparticle (or antimatter counterpart, i.e. antielectron) of the electron. It is generated in electron-positron pair creation by gammas interacting with nuclei and it is emitted during beta plus decay.

Ppm (parts per million, 1 part per 1 million parts) Measure of the degree of impurities in solid bodies, liquids and gases.

Pressure tube reactor Nuclear reactor in which the fuel elements are contained within many tubes through which the coolant is circulated. The moderator surrounds this tube arrangement. Examples of this type of reactors are CANDU and RBMK reactors.

Pressure vessel Thick-walled cylindrical steel vessel enclosing the reactor core in a nuclear power plant. The vessel is made of a special fine-grained steel well suited for welding and with a high toughness while showing low porosity under neutron irradiation.

Pressurised water reactor (PWR) Common nuclear power reactor design in which very pure water is maintained under a high pressure to keep it from boiling at the high operating temperature. The thermal energy generated by the reactor is transferred from the core to a large heat exchanger that heats water in a secondary circuit to produce the steam needed to generate electricity.

Probabilistic safety assessment (PSA) Safety analysis that uses probabilistic risk assessment techniques during both the design and operation of a nuclear power plant to analyse the overall risk. Considering an entire set of potential events with their respective probabilities and consequences, the overall risk of a nuclear incident or accident can be assessed.

Proton (symbol p) Elementary nuclear particle with a positive electric charge and a mass of 1.67262×10^{-27} kg. It is present in the nucleus of an atom.

Proton emission One process that unstable atoms can use to become more stable. During proton emission, a proton is ejected from an atomic nucleus. Since an atom loses a proton during proton emission, it changes from one element to another. For example, after undergoing proton emission, an atom of nitrogen (with 7 protons) becomes an atom of carbon (with 6 protons).

Radiation Energy travelling in the form of high-speed particles or electromagnetic waves.

Radioactivity The spontaneous change of an unstable nucleus into one or more different nuclei, often resulting in the emission of radiation. This process is referred to as a transformation, a decay, or a disintegration of an atom. Radioactive atoms are often called radioactive isotopes or radionuclides. This change occurs over a characteristic period of time, called a *half-life*.

Radioactive series Succession of nuclides, each of which transforms by radioactive disintegration into the next until a stable nuclide results. The first member is called the parent, the intermediate members are called daughters, and the final stable member is called the end product.

Radiotherapy Any radiation treatment of human beings with ionizing radiation. Radiation treatments are typically carried out to treat cancer.

Radiotoxicity Measure of the harmfulness of a radionuclide. The type and energy of rays, absorption in the organism, residence time in the body, etc. influence the degree of radiotoxicity of a radionuclide.

Radium (symbol Ra) Radioactive metallic element with atomic number 88. As found in nature, the most common isotope has a mass number of 226. It occurs in minute quantities associated with uranium in pitchblende, camotite, and other minerals.

Radon (symbol Rn) Radioactive element with atomic number is 86. It is one of the heaviest gases known. It is a daughter of radium.

Reactivity The reactivity of a nuclear reactor expresses the departure of that reactor from criticality, the normal operating condition of a reactor, in which nuclear fuel sustain a steady-state fission chain reaction. A positive reactivity addition indicates a move towards supercriticality (power increase). A negative reactivity addition indicates a move towards subcriticality (power decrease). Control rods are the main reactivity control systems.

Reactor (see Nuclear Reactor).

Reactor core (see Core).

Reactor-years Measure of reactor experience. A reactor-year is one year of operation of a single reactor.

Recycling (see Reprocessing).

Reflector Material surrounding the reactor core, with the property of scattering back into the core a certain amount of neutrons that would escape without it. Its function is to decrease the leakage factor, thereby increasing the multiplication coefficient of the chain reaction. Common reflector materials are graphite, beryllium, water, stainless still and natural uranium.

Reprocessing (recycling) Process in which spent fuel is treated to recover the fissile isotopes of uranium and plutonium, by separating them from fission products and minor actinides. The fissile isotopes are then used to produce fresh fuel called MOX (Mixed-OXide). It provides a more efficient use of the fuel, thereby reducing the amount of waste to be disposed of.

Risk-informed regulation Integration of quantitative and qualitative, deterministic and probabilistic safety analyses to determine aspects of plant design and operation. It takes into account event likelihood, potential consequences, good practices and sound management.

Safeguards Material control and accounting programs to verify that all special nuclear material is properly controlled and accounted for, as well as to verify the physical protection (or physical security) equipment and security forces. As used by the International Atomic Energy Agency, this term also means verifying that the peaceful use commitments made in binding non-proliferation agreements are honoured.

Safety culture Set of characteristics and attitudes in organisations and individuals which establishes that, as an overriding priority, nuclear plant safety issues receive the attention warranted by their significance, to ensure the protection of people and the environment.

Scattering Collision of an incident nucleus/particle with a target nucleus in which the incident projectile is found among the products of the reaction.

Scram Term used to describe the sudden shutting down of a nuclear reactor. It is believed to have been originally an acronym meaning "safety control rod axe man" used at the first operating reactor in the United States, the Chicago Pile.

Severe accident In a nuclear reactor, it is an event which significantly exceeds design basis events and conditions, and is characterised by extensive core damage (molten core) due to an excessive reactivity excursion or the inability to provide adequate cooling to the core.

Shielding Any material or obstruction that absorbs radiation and thus can protect personnel or materials from the effects of ionizing radiation.

Sievert (Sv) It is the international system (SI) unit for dose equivalent, equal to 1 Joule per kilogram. Named after the Swedish medical physicist Rolf M. Sievert.

Small modular reactor (SMR) A new generation of advanced reactors typically in the range of 50–300 MW, characterised by modular design and construction to be built in factories and shipped to utilities for installation as demand arises.

Spent nuclear fuel (SNF) Fuel that has been irradiated in and then permanently removed from a nuclear reactor.

Steam generator The heat exchanger used in some reactor designs to transfer heat from the primary (reactor coolant) system to the secondary (steam) system. This design permits heat exchange with little or no contamination of the secondary system equipment.

Stochastic effects Effects (e.g. cancer, leukaemia, and genetic effects) that occur by chance, whose probability of occurring is assumed to be proportional to the radiation exposure received.

Subcritical mass Amount of fissile material insufficient in quantity or of improper geometrical configuration to sustain a fission chain reaction.

Subcriticality The condition of a reactor core, in which on average less than one neutron per fission is available to produce yet another fission, so that the chain reaction cannot propagate.

Supercritical reactor Nuclear reactor in which on average more than one neutron per fission is available to produce yet another fission, so that the chain reaction has a diverging character and the power level is increasing with time.

Supercriticality The condition in which the rate of fission neutron production exceeds all neutron losses, and the overall neutron population increases, as does the reactor power.

Terrestrial radiation The portion of the natural background radiation that comes from the Earth itself and is produced by the decay of primordial and cosmogenic radionuclides such as uranium, thorium, and radon in the earth.

Thermal neutrons (see neutron, thermal).

Thermal reactor Reactor in which the fission chain reaction is sustained by thermal neutrons. Most current reactors are thermal reactors.

Thermal reactor recycle Reprocessing and recycle of plutonium (and uranium) in thermal reactors.

Thorium (symbol Th) Weakly radioactive, silvery metallic element with atomic number 90. Thorium-232, which has 142 neutrons, is the most stable isotope of thorium and accounts for nearly all natural thorium, with the other five natural isotopes occurring only in traces: it decays very slowly through alpha decay to radium-228, starting a decay chain named the thorium series that ends with lead-208. Thorium is estimated to be about three times more abundant than uranium in the Earth's crust, and is chiefly refined from monazite sands as a by-product of extracting rare earth metals.

Thorium fuel cycle Cycle in which fertile Th-232 is converted into fissile U-233.

Transmutation Nuclear reaction where there is a rearrangement of nuclear constituents between the colliding nuclei. A typical example is the neutron capture reaction in which the target nucleus absorbs a neutron and changes into another isotope. This process occurs in fission reactors and is the process by which some long-lived elements of radioactive waste are created. It is also a process being investigated as a means to transform long-lived elements of high-level radioactive waste into shorter-lived elements.

Transuranic element An artificially made, radioactive element that has an atomic number Z higher than that of uranium ($Z = 92$), such as neptunium ($Z = 93$), plutonium ($Z = 94$), americium ($Z = 95$), and others.

Tritium (symbol ^{3}H or T) Radioactive isotope of hydrogen having two neutrons and one proton in the nucleus. It decays by emitting beta particles and has a half-life of about 12.32 years.

Triton (symbol t) Nucleus of tritium. It contains one proton and two neutrons.

UNSCEAR (United Nations Scientific Committee on the Effects of Atomic Radiation) Scientific Committee established by the General Assembly of the United Nations in 1955 with the mandate to assess and report levels and effects of exposure to ionising radiation. It is based in Wien, Austria (http://www.unscear.org/). Governments and organizations throughout the world rely on the Committee's estimates as the scientific basis for evaluating radiation risk and for establishing protective measures.

Uranium (symbol U) Radioactive element with the atomic number $Z = 92$. It is found in natural ores as a mixture of the three isotopes ^{238}U (99.275 %), ^{235}U (0.720 %) and ^{234}U (0.005 %), where numbers in parentheses show the relative percentages. Uranium-235 is fissile, while uranium-238 is fissionable by fast neutrons and is fertile, meaning that it becomes fissile after absorbing one neutron and after undergoing a few radioactive decays.

Vitrification Process of producing glass. It is a technology commonly used to immobilise the high-level waste produced from the reprocessing of spent nuclear fuel. Typically this glass features high durability, the ability to withstand the intense radiation and high heat associated with high-level waste and the stability necessary to contain the radioactive isotopes over long periods of time.

Waste (radioactive) The term radioactive waste indicates all residual materials with some degree of radioactivity above the natural background, arising from any type of nuclear activity. The name waste implies that no further use of such materials is possible so that they have to be disposed of as it happens with other, non-radioactive waste.

Waste transmutation Reactor transmutation of long-lived fission products or actinides to stable elements or those that are less radiotoxic.

Watt It is the international system (SI) unit for power, defined as the consumption or conversion of one joule of energy per second (1 W = 1 J/s).

Watthour Unit of energy equal to one watt of power generated continuously by a system for one hour: 1 Wh = 3.6×10^3 J.

WANO (World Association of Nuclear Operators) International organisation that unites every company and country in the world hosting operating commercial nuclear power plants, with the purpose to achieve the highest possible standards of nuclear safety (http://www.wano.info/en-gb/).

WNA (World Nuclear Association) International organisation that represents the global nuclear industry (http://world-nuclear.org/). Its mission is to promote a wider understanding of nuclear energy among key international influencers by producing authoritative information, developing common industry positions, and contributing to the energy debate.

X-ray X-rays are electromagnetic waves emitted by energy changes in an atom's electrons. Their wavelength is much shorter than that of visible light. They are a form of high-energy electromagnetic radiation that interacts lightly with matter and can ionize atoms and molecules. Thick layers of lead or other dense materials are suitable to stop them.

Yellow cake The solid form of mixed uranium oxide, which is produced from uranium ore in the uranium recovery (milling) process. The material is a mixture of uranium oxides (UO_2 and UO_3 in 1 to 2 ratio), known as pitchblende, commonly referred to as U_3O_8.

Printed in the United States
By Bookmasters